Thomas Barth
Anke Schüll (Hrsg.)

Grid Computing

Delivering Security for the Future

EADS Military Aircraft is an international systems integrator, and a total service provider working today for tomorrow's future.

We design, integrate and deliver manned and unmanned airborne systems which Air Forces require for combat, reconnaissance and training.

Dedicated to our customers needs, we can provide complex aerostructures and also offer logistic support for high-performance fighters, transport and mission aircraft.

EADS

Military Aircraft, Communication

81663 Munich · Germany
Phone: +49 (0) 89. 6 07-2 57 11
Telefax: +49 (0) 89. 6 07-2 24 55

EADS DEFENCE & SECURITY

Thomas Barth
Anke Schüll (Hrsg.)

Grid Computing

Konzepte – Technologien –
Anwendungen

Mit 60 Abbildungen

Bibliografische Information Der Deutschen Nationalbibliothek
Die Deutsche Nationalbibliothek verzeichnet diese Publikation in der
Deutschen Nationalbibliografie; detaillierte bibliografische Daten sind im Internet über
<http://dnb.d-nb.de> abrufbar.

Die Wiedergabe von Gebrauchsnamen, Handelsnamen, Warenbezeichnungen usw. in diesem Werk berechtigt auch ohne besondere Kennzeichnung nicht zu der Annahme, dass solche Namen im Sinne von Warenzeichen- und Markenschutz-Gesetzgebung als frei zu betrachten wären und daher von jedermann benutzt werden dürfen.

Höchste inhaltliche und technische Qualität unserer Produkte ist unser Ziel. Bei der Produktion und Auslieferung unserer Bücher wollen wir die Umwelt schonen: Dieses Buch ist auf säurefreiem und chlorfrei gebleichtem Papier gedruckt. Die Einschweißfolie besteht aus Polyäthylen und damit aus organischen Grundstoffen, die weder bei der Herstellung noch bei der Verbrennung Schadstoffe freisetzen.

1. Auflage März 2006

Alle Rechte vorbehalten
© Friedr. Vieweg & Sohn Verlag | GWV Fachverlage GmbH, Wiesbaden 2006

Lektorat: Dr. Reinald Klockenbusch / Andrea Broßler

Der Vieweg Verlag ist ein Unternehmen von Springer Science+Business Media.
www.vieweg.de

Das Werk einschließlich aller seiner Teile ist urheberrechtlich geschützt. Jede Verwertung außerhalb der engen Grenzen des Urheberrechtsgesetzes ist ohne Zustimmung des Verlags unzulässig und strafbar. Das gilt insbesondere für Vervielfältigungen, Übersetzungen, Mikroverfilmungen und die Einspeicherung und Verarbeitung in elektronischen Systemen.

Konzeption und Layout des Umschlags: Ulrike Weigel, www.CorporateDesignGroup.de
Umschlagbild: Nina Faber de.sign, Wiesbaden
Gedruckt auf säurefreiem und chlorfrei gebleichtem Papier.
Printed in Germany

ISBN 978-3-8348-0033-6

Vorwort

Grid Computing als Forschungsrichtung der Informatik und zunehmend auch der Wirtschaftsinformatik vereinigt viele unterschiedliche Problemfelder, die eine erhebliche Bandbreite von ökonomischen Modellen der Nutzung von IT-Funktionalitäten bis hin zur technischen Umsetzung verteilter Systeme aufweisen. Im Vergleich zu den Arbeiten hauptsächlich der Informatik, die sich bereits seit mehreren Jahrzehnten sehr intensiv dem Entwurf, der Implementierung und der Nutzung verteilter Systeme widmen, zeichnete sich der Ansatz des Grid Computing bereits von Beginn an dadurch aus, verteilte Systeme nicht schwerpunktmäßig als Instrument zur Bearbeitung wissenschaftlicher Probleme mit Mitteln des Höchstleistungsrechnens zu begreifen und darzustellen. Vielmehr wurde bereits zu Beginn der Arbeiten auf dem Gebiet des Grid Computing Mitte der 90er Jahre des 20. Jahrhunderts ein Schwerpunkt darauf gelegt, die durch eine Grid Infrastruktur zusammengefassten, typischerweise heterogenen und verteilten Ressourcen so für Anwender verfügbar zu machen, dass auch in kommerziellen, wettbewerblichen Szenarien eine Nutzung in Frage kommt. Dieses Ziel machte die Integration von Funktionalitäten wie Lizenzmanagement, Sicherheit sowie Leistungsverrechnung von Ressourcennutzung notwendig. Diese Aspekte sind in einem rein wissenschaftlichen Umfeld von geringerer Bedeutung und damit auch nicht Gegenstand der Forschung.

Einen Ausschnitt der hier skizzierten Anforderungen an Grid Computing-Infrastrukturen decken die in diesem Band zusammengefassten Beiträge ab, ohne jedoch Anspruch auf Vollständigkeit erheben zu wollen.

Im einleitenden Beitrag werden von Harms, Rehm, Rueter und Wittmann anhand konkreter Projekte der aktuelle Stand von Grid Computing-Lösungen sowie die Entwicklung hin zu virtualisierten IT-Infrastrukturen dargestellt. Hierbei wird die Motivation der Kunden ebenso wie der Anbieter von Grid Computing-Lösungen deutlich. Ebenfalls aus der Perspektive eines Anbieters von Grid Computing-Lösungen wird im zweiten Kapitel durch Geiger ein Schwerpunkt auf Services als die grundlegenden Elemente einer zukünftigen IT-Infrastruktur gelegt. Der aus der zentralen Bedeutung von Services resultierenden Fragestellung der Auswahl des „optimalen" Services unter ökonomischen Qualitäts-Kriterien in einem Netzwerk widmen sich Eymann, Reinicke und Streitberger in Kapitel drei. Im daran anschließenden Kapitel von Neumann, Veit und Weinhardt steht die Übertragung von Marktmechanismen auf einen Grid-basierten Markt für Grid-Ressourcen im Mittelpunkt.

Nach diesen Kapiteln, welche die strategische Bedeutung von Grid Computing-Ansätzen sowie einige grundlegende Konzepte beim Agieren auf einem Ressourcen-Markt thematisieren, wird durch Resch die Weiterentwicklung des Grid-Ansatzes zu einer umfassenden „Workbench" für die verteilte Kooperation beschrieben. Der Einsatz rechenzeitintensiver Simulationen in Wissenschaft und Industrie bildet hierbei die Verbindung zu den folgenden Kapiteln, in denen ein Überblick über die spezifischen Anforderungen an Grid-Systeme aus den Blickwinkeln so unterschiedlicher Anwendungsbereiche wie der „Systems Biology" (im Beitrag von Wiechert, Haunschild, Weitzel, Nöh, von Lieres, Wahl, Qeli und Freisleben) und der Gießereitechnik (durch Jakumeit in Kapitel sieben) gegeben wird.

In den abschließenden Kapiteln werden durch Friese, Smith und Freisleben zwei innovative technische Ansätze dargestellt, durch die der Einsatz der Grid Computing-Konzepte in globalen Szenarien ermöglicht wird: Für das service-orientierte Ad Hoc-Grid (Kapitel acht) wird dabei die verbreitete Analogie mit dem Stromnetz aufgegriffen, welche die einfache und intuitive Benutzbarkeit von Ressourcen überall verdeutlicht und in diesem Kontext die Ausbreitung und Nutzbarkeit von Services ohne expliziten zusätzlichen Administrationsaufwand anstrebt. Im zweiten Beitrag wird die modell-getriebene Generierung von Services thematisiert, die die Lücke zwischen anwendungs-spezifischen Abläufen und service-orientiertem Code schließen soll.

Dieser Band entstand anlässlich des 60-ten Geburtstages von Prof. Dr. Manfred Grauer, der in seinen Arbeiten am Institut für Wirtschaftsinformatik der Universität Siegen die Potentiale des Grid Computing insbesondere für die mittelständische Industrie und ihren komplexen, ingenieurwissenschaftlichen Fragestellungen frühzeitig erkannte. In einer Vielzahl von Kooperationen mit der Industrie und wissenschaftlichen Veröffentlichungen sind die Spezifika des Einsatzes von Grid Computing-Konzepten im kommerziellen Umfeld erarbeitet und Lösungsansätze erforscht und umgesetzt worden. Durch die Initiierung und Mitwirkung an Projekten auf Landes-, Bundes- und europäischer Ebene konnten die Ergebnisse dieser Arbeiten, beispielsweise im Rahmen der bundesweiten D-Grid-Initiative des Bundesministerium für Bildung und Forschung (www.d-grid.de), in den Aufbau einer deutschen Grid Infrastruktur einfließen. Unter den Autoren, die zu diesem Band beigetragen haben, finden sich einige der akademischen und industriellen Kooperationspartner, die gemeinsam diese Projekte ermöglicht und durchgeführt haben bzw. an ihnen mitwirken.

Mit diesem Buch möchten wir uns bei Prof. Dr. Manfred Grauer für seine vielfältige Unterstützung in den vergangenen Jahren bedanken. Seine fachliche Begleitung sowie sein breit gefächertes Engagement haben zu vielen Denkanstößen geführt und unsere akademischen Laufbahnen entscheidend geprägt.

Ebenso möchten wir uns natürlich bei den Autoren bedanken, den Kolleginnen und Kollegen des Instituts für Wirtschaftsinformatik der Universität Siegen sowie einer ganzen Reihe von aktiven und interessierten Unternehmen, die in unterschiedlicher Art und Weise einen Beitrag dazu geleistet haben, dieses Buch zu ermöglichen. Eine vollständige Aufzählung würde diesen Rahmen sprengen und wir verweisen stattdessen auf www-winfo.uni-siegen.de/GridComputingBuch. Besonderer Dank gilt Frau Annette Wiebusch sowie den Herren Markus Hoffmann und Gregor Stuhldreier, die bei der Vorbereitung des Manuskriptes einen wertvollen Beitrag geleistet haben.

Während der Arbeit an einem Beitrag zu diesem Buch verstarb zu unserem großen Bedauern Herr Uwe Harms, ein langjähriger und ausgewiesener Experte auf den Gebieten Supercomputing und Grid Computing. Wir möchten auch an dieser Stelle an ihn und seine Arbeiten erinnern.

Wir hoffen, mit dem vorliegenden Band eine Übersicht über die Vielschichtigkeit dieses Gebietes geben zu können. Anhand der hier dargestellten grundlegenden Konzepte, Anwendungs-Szenarien und den Ansätzen zur Realisierung von Grid-Infrastrukturen möchten wir sowohl Wissenschaftlern als auch IT-Verantwortlichen unterschiedliche Aspekte und Potentiale vermitteln, welche die Bedeutung des Grid Computing für leistungs- und anpassungsfähige zukünftige IT-Infrastrukturen ausmachen.

Siegen, im Januar 2006

Thomas Barth   Anke Schüll

Inhaltsverzeichnis

1 Grid Computing für virtualisierte Infrastrukturen 1
   1.1 Grid Computing-Überblick (von Uwe Harms) 1
   1.2 IBM Grid Implementierungen 3
   1.3 Schwerpunkt des Grid Computing bei IBM 9
   1.4 Grid-Infrastrukturen der Zukunft 12
   1.5 Grid und Virtualisierung 15
   1.6 Literaturverzeichnis 15

2 Service Grids – von der Vision zur Realität 17
   2.1 Einführung 17
   2.2 eScience und eEngineering im Service-Grid 18
   2.3 Die Basis: Grid-Technologie 20
      2.3.1 Internet – Web – Grid 20
      2.3.2 Konvergenz von IT und TC 21
      2.3.3 Ausprägungen von Grids 21
   2.4 Architektur 24
   2.5 Auswirkungen 27
   2.6 Status quo: Application-Grid für IT-Services im DLR 30
   2.7 Ein Modell der zweiten Generation für ASP 31
   2.8 Zusammenfassung und Ausblick 32

3 Ökonomische Bewertung der Dienstauswahlverfahren in Service-Netzen 33
   3.1 On-demand Computing in der serviceorientierten Architektur (SOA) 34
      3.1.1 On-demand Computing: Kostensenkung oder Risikoerhöhung? 34
      3.1.2 Unterstützung der Transaktionsphasen durch Standardisierung 36
      3.1.3 Technische und ökonomische Bewertungsmetriken 37

	3.2	Dienstauswahlverfahren zur effektiven Selektion	40
	3.2.1	Zentrale Dienstfindung und -auswahl	42
	3.2.2	Auswahl der Dienste beim Konsumenten (dezentrale Dienstauswahl)	45
	3.2.3	Ordnungsmethoden	45
3.3		Simulation und Evaluation	48
	3.3.1	Netzwerkattribute und Hypothesen	49
	3.3.2	Der Netzwerksimulator J-Sim/TCL	51
	3.3.3	Simulationsergebnisse	52
	3.3.4	Synopse	58
3.4		Fazit und Ausblick	58
3.5		Literaturverzeichnis	59
4		**Grid Economics: Market Mechanisms for Grid Markets**	**64**
4.1		Introduction	64
4.2		The Grid Environment and Corresponding Requirements	66
	4.2.1	The Grid Environment	67
	4.2.2	Requirements	68
	4.2.2.1	Requirements on the outcome	68
	4.2.2.2	Requirements upon the mechanism	70
4.3		Market Mechanisms for Grid	72
	4.3.1	Classic Auction Types	72
	4.3.2	Combinatorial Auctions and Exchanges	74
	4.3.3	Scheduling Auctions	76
4.4		Impediments of Market Mechanism Adoption	77
4.5		Concluding Remarks	79
4.6		References	80
5		**Grid at the Interface of Industry and Research**	**85**
5.1		What is a Grid?	85
5.2		First approach to Grid Computing	86
5.3		The HLRS Teraflop Workbench Concept	87
5.4		Steps of an Integrated Simulation Workflow	88
	5.4.1	Process Chain Integration	88

	5.4.2	Simulation Steps	89
	5.5	Requirements	91
	5.5.1	Data	91
	5.5.2	Networks	91
	5.5.3	Software	92
	5.5.4	Summary of Requirements	93
	5.6	Concept	93
	5.6.1	File System	93
	5.6.2.	Integration of Heterogeneous Servers	94
	5.6.3.	Integration of Visualization and Supercomputing	94
	5.6.4	Software Integration	95
	5.7	Conclusion	95
	5.8	References	96
6	**Grid Computing for Systems Biology**		98
	6.1	Introduction	99
	6.1.1	Systems Biology	99
	6.1.2	Systems Thinking	101
	6.1.3	Central Problems of Biological Systems Modeling	102
	6.1.4	Model Guided Discovery	104
	6.2	Grid Computing	106
	6.2.1	Using Grid Resources for Biological Research	106
	6.2.2	Current Developments	108
	6.3	^{13}C Metabolic Flux Analysis	111
	6.3.1	Simulation and Sensitivity Analysis	111
	6.3.2	High-Throughput Flux Analysis	112
	6.3.3	Nonlinear Error Propagation	113
	6.3.4	Isotopically Instationary Experiments	114
	6.3.5	Optimum Experimental Design	116
	6.4	Evaluation of Stimulus Response Experiments	117
	6.4.1	Simulation and Sensitivity Analysis	117
	6.4.2	Metabolic Modeling Tool	119

	6.4.3	External Inputs 120
	6.4.4	Model Selection 121
	6.4.5	Technical Considerations of Grid Implementation 123
6.5	Conclusion... 125	
6.6	References 126	

7 Grid-basierte Simulation für die Gießerei-Industrie 134

7.1	Simulation in der Gießerei-Industrie 135
7.2	Szenarien für den Einsatz von Grid-Technologie 137
7.3	Beispiel aus der Praxis.. 141
7.4	Zusammenfassung... 142
7.5	Literaturverzeichnis 143

8 Service-Oriented Ad Hoc Grids.. 144

8.1	Introduction.. 144	
8.2	The Ad Hoc Grid 146	
8.3	Challenges 147	
	8.3.1	Node Communication 147
	8.3.2	Node / Service Discovery..................................... 148
	8.3.3	Service Deployment and Administration 149
	8.3.4	Service Security... 149
	8.3.5	Service Trust 150
8.4	Related Work.. 151	
8.5	Implementation of a Service-Oriented Ad Hoc Grid .. 154	
	8.5.1	P2P Infrastructure 154
	8.5.2	Node Discovery 157
	8.5.3	Service Discovery 159
	8.5.4	Service Invocation 160
	8.5.5	Service Deployment and Administration 164
	8.5.6	Service Security... 169
	8.5.7	Service Trust 179
8.6	Conclusions 187	
8.7	Acknowledgements.. 188	
8.8	References 188	

9		Model Driven Development of Service-Oriented Grid Applications	193
	9.1	Introduction	193
	9.2	Related Work	196
	9.2.1	Service-Oriented Grid Computing	196
	9.2.2	Model Driven Architecture	196
	9.3	MDA Meets the Grid: An Application Example	198
	9.3.1	PIM Layer: Business View	198
	9.3.2	PSM Layer: Grid Service Design	199
	9.3.3	Code Layer: Grid Service Implementation	201
	9.3.4	Separation of Concerns	203
	9.4	An MDA Approach to Service-Oriented Grid Computing	203
	9.4.1	PSM Layer: Grid Profile	203
	9.4.2	Code Layer: Java Annotations	207
	9.5	Conclusions	210
	9.6	Acknowledgements	210
	9.7	References	211
Index			213

Autorenverzeichnis

Prof. Dr. T. Eymann	Universität Bayreuth Wirtschaftsinformatik (BWL VII)
Prof. Dr. B. Freisleben	Philipps-Universität Marburg, Fachbereich Mathematik und Informatik, Verteilte Systeme
Dipl.-Inform. Th. Friese	Philipps-Universität Marburg, Fachbereich Mathematik und Informatik, Verteilte Systeme
Priv.-Doz. Dr. A. Geiger	T-Systems Solutions for Research GmbH, Weßling
U. Harms †	Freier Autor (für IBM Deutschland)
Dipl.-Ing. M. D. Haunschild	Universität Siegen, Fachbereich Maschinenbau, Lehrstuhl für Simulationstechnik
Priv.-Doz. Dr. J. Jakumeit	ACCESS e.V., Aachen
E. von Lieres	Universität Siegen, Fachbereich Maschinenbau, Lehrstuhl für Simulationstechnik
Dr. D. Neumann	Universität Karlsruhe (TH) Fakultät für Wirtschaftswissenschaften Lehrstuhl für Informationsbetriebswirtschaftslehre
Dipl.-Math K. Nöh	Universität Siegen, Fachbereich Maschinenbau, Lehrstuhl für Simulationstechnik
Dipl.-Inform. E. Qeli	Universität Siegen, Fachbereich Maschinenbau, Lehrstuhl für Simulationstechnik
H.-J. Rehm	Kommunikation Systems & Technology Group, IBM Deutschland

Dipl.-Kfm. Univ. M. Reinicke	Universität Bayreuth, Wirtschaftsinformatik (BWL VII)
Prof. Dr. M. Resch	Universität Stuttgart, Institut für Höchstleistungsrechnen und Höchstleistungsrechenzentrum Stuttgart (HLRS)
Th. Rüter	IBM Grid Enablement Manager, IBM Deutschland
M. Smith	Philipps-Universität Marburg, Fachbereich Mathematik und Informatik, Verteilte Systeme
Dipl.-Inform. W. Streitberger	Universität Bayreuth, Wirtschaftsinformatik (BWL VII)
Dr. D. Veit	Universität Karlsruhe (TH) Fakultät für Wirtschaftswissenschaften Lehrstuhl für Informationsbetriebswirtschaftslehre
Dipl.-Ing. A. Wahl	Universität Siegen, Fachbereich Maschinenbau, Lehrstuhl für Simulationstechnik
Prof. Dr. Chr. Weinhardt	Universität Karlsruhe (TH) Fakultät für Wirtschaftswissenschaften Lehrstuhl für Informationsbetriebswirtschaftslehre
Dipl.-Ing. M. Weitzel	Universität Siegen, Lehrstuhl für Simulationstechnik
Prof. Dr. W. Wiechert	Universität Siegen, Fachbereich Maschinenbau, Lehrstuhl für Simulationstechnik
H. M. Wittmann	Technical Director Systems & Technology Group, IBM Deutschland

1 Grid Computing für virtualisierte Infrastrukturen

U. Harms †, H.-J. Rehm, T. Rueter und H. Wittmann

Die Veröffentlichungen von IBM zum Thema Grid Computing sind sehr eng mit dem Namen Uwe Harms verbunden. Uwe Harms hat mit kritischem, aber stets fairem Auge als intimer Kenner der Supercomputing-Center die Aktivitäten der IBM in Umfeld der IBM RS/6000 und der pSeries begleitet.

Uwe Harms hat Teile dieses Artikels als eines seiner letzten Projekte für uns bearbeitet. Alle, die wir bei IBM mit ihm in langjährigem Kontakt standen, haben uns stets über seine freundliche, faire und ehrliche Zusammenarbeit gefreut, und werden die Erinnerung an ihn bewahren.

1.1 Grid Computing-Überblick (von Uwe Harms)

Die Begriffe Grid bzw. Grid Technologie oder Grid Computing leiten sich vom englischen Begriff für das Stromnetz („power grid") ab, durch das jeder angeschlossene Nutzer eine „Leistung" – hier: elektrischen Strom – auf einfache Art „beziehen" kann, ohne die vollständige Infrastruktur zur Stromerzeugung und – weiterleitung besitzen zu müssen: Man schaltet morgens die Kaffeemaschine an und sie arbeitet. Als Nutzer weiß man nicht, woher der Strom kommt und wie er produziert wurde. Diese Aufgabe regeln im Hintergrund die Stromversorger unter sich. Mit den Stadtwerken beispielsweise hat der Kunde seinen Vertrag und zahlt seine Grund- und Verbrauchsgebühr.

Die Europäische Gemeinschaft investierte im 6. Rahmenprogramm für 2002-2006 (www.rp6.de) 125 Millionen Euro in das Thema Grid als Forschungsauftrag und die gleiche Summe für das Installieren von Grid-basierten Infrastrukturen für die Forschung in Europa. Damit werden beispielsweise die europäischen Supercomputerzentren miteinander verbunden und können ihre Ressourcen teilen. Und wie in jedem Ansatz von Verteilung von Ressourcen gibt es auch hier verwaltungstechnische und gesetzliche Anforderungen zu erfüllen. Darf ein französischer Forscher einen Rechner nutzen, der mit deutschen Steuergeldern finanziert wurde? Diese Grid-Ideen sind für den Mittelständler heutzutage noch illusorisch.

Einige Grundideen des Grid Computing wurden schon 1992 unter anderen Bezeichnungen aufgegriffen: Load Sharing oder Lastverteilung etwa. Hierbei engagierten sich zwei Unternehmen besonders: Genias mit „Codine" adressierte mehr den akademischen Markt, Platform Computing (www.platform.com) mit der „Load Sharing Facility" (LSF, [1]) konzentrierte sich auf das industrielle Umfeld. Sun Microsystems (www.sun.com) kaufte im Jahre 2000 den Nachfolger von Genias, „Gridware", und bietet Teile der Software kostenlos als „Sun Grid Engine" [2] an.

Die Motivation für die Nutzung dieser Ideen lässt sich am besten anhand eines typischen Szenarios darstellen: Pragmatisch ging 1997 die DASA Airbus (www.airbus.dasa.de) die Nutzung der Ideen des Grid Computing an. Die Entwurfsworkstations der Ingenieure wurden tagsüber nur zu 10% genutzt und standen am Wochenende und über Nacht ungenutzt im Werk. Mit Hilfe von LSF konnten die Ingenieure rechenzeitintensive Simulationsrechnungen an eine Middleware übergeben, welche die Verteilung dieser Jobs vornahm. Diese suchte die passenden leistungsfähigen Knoten des Netzwerks aus, auf der auch die geforderte Software verfügbar war. Wenn die ausgewählte Workstation frei war, startete LSF den Job, der dann über Nacht oder das Wochenende ausgeführt wurde. Ohne zusätzliche Investitionen - außer für LSF - ließ sich auf diese Art vorhandene und ungenutzte Rechenleistung abrufen. Die Auslastung der Workstations wuchs dadurch erheblich und es ließ sich eine Kostenersparnis realisieren, da keine zusätzlichen Rechner für die Deckung des Bedarfs an Rechenleistung beschafft werden mussten.

Ein derartiger Ansatz ist sicherlich in vielen Szenarien relevant – etwa für mittelständische Automobilzulieferer oder Ingenieurbüros –, in denen vorhandene Rechenleistung ungenutzt ist.

1.2 IBM Grid Implementierungen

IBM hat mit seinen Kunden seit Gründung der Grid Initiative innerhalb der IBM zahlreiche Referenzen im deutschsprachigen Raum aufbauen können. So wurden hier die ersten Grid Referenzen im industriellen Umfeld mit der Firma Magna Steyr und Siemens Mobile aufgebaut. Bei diesen Projekten stand der Aspekt der optimalen Nutzung der Rechen-Ressourcen für verteilte Anwendungen im Vordergrund. Interessant ist hierbei, dass diese Projekte gut mit der zu der damaligen Zeit verfügbaren Grid-Software (in diesen Fällen wiederum LSF) durchgeführt werden konnten. Bei beiden Projekten war die eigentliche Aufgabe das Aufbereiten der Daten beziehungsweise die Änderung der Arbeitsprozesse und ihre IT-Implementierung, das als signifikanter Fortschritt herausgestellt werden kann.

Im Falle des Projekts bei Magna Steyr wurde eine Maßüberschneidungsanalyse (Clash-Analyse) von einer einzigen AIX Workstation in ein Netz vieler heterogener Workstation verteilt (zu weiteren Informationen über dieses Projekt s. [3]). Hier stand im technischen Vordergrund des Projekts das Erstellen einer Verteilungs- und Sammlungsfunktion.

1 Grid Computing für virtualisierte Infrastrukturen

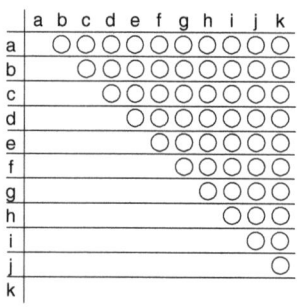
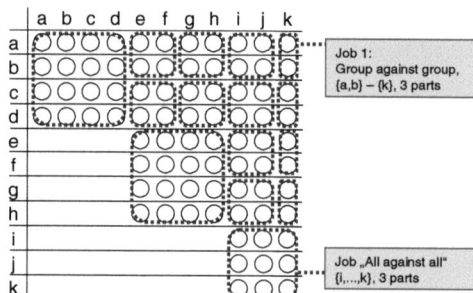

Abb. 1: *Maßüberschneidungsanalyse bei Magna Steyr*

Es war zu berücksichtigen, dass die Maßüberschneidungsrechnung grundsätzlich eine symmetrische Berechnung ist (s. Abb. 1). Eine Überschneidung zwischen den Elementen *a* und *b* ist gleich der Überschneidung der Elemente *b* und *a*. Das Grid-System hat hier dann die Aufgabe, eine optimale Verteilung der Berechnungen auf die bereitstehenden heterogenen Ressourcen vorzunehmen. Die Aufgabenpakete werden verteilt berechnet und von einer zentralen Instanz durch eine Endanalyse konsolidiert. Da in diesem Projekt das CAD/CAE-System CATIA von Dassault (www.dassault.com) zum Einsatz kam, konnten hier nur mit der Hilfe von Dassault selbst an den internen Schnittstellen die notwendigen Ergänzungen vorgenommen werden, um die beschriebene verteilte Verarbeitung zu realisieren.

Die beiden größten Herausforderungen bei rechenzeitintensiven und komplexen Aufgaben wie der Maßüberschneidungsanalyse Maßüberschneidungsanalyse sind, dass auf der einen Seite der Benutzer einen zu hohen administrativen Aufwand hat, die Analyse-Software zu benutzen, und auf der anderen Seite zu viel Zeit

1.2 IBM Grid Implementierungen

vergeht, bis das Ergebnis einer Analyse zurückgegeben wird. Da das Modell (und damit die Anzahl Elemente) stetig wächst, sind überproportional mehr Überprüfungen aller Element untereinander nötig, was die Berechungszeit stetig erhöht und letztendlich die Zeit, bis das Produkt am Markt verfügbar ist („time to market").

Durch die Verwendung des Grid-Ansatzes konnte die Berechnung einer Maßüberschneidungsanalyse signifikant in der Laufzeit reduziert werden, so dass die Mitarbeiter nun in der Lage waren, die Analysen mehrmals am Tag auszuführen, was früher undenkbar war. So können mehr Design-Alternativen in Betracht gezogen werden, da diese von System schneller neu berechnet werden konnten und Fehler im Modell frühzeitig erkannt wurden.

Siemens Mobile in Kamp-Lintfort benötigte für den ständig wachsenden Bedarf an Rechenleistung bei der Software-Entwicklung für Mobile Phones eine verlässliche skalierbare Platform, um ein zentrales System, dass neben dem zentralen, notwendigen Current Versioning System (CVS) auch noch die Entwicklungsumgebung der Softwareingenieure trug, zu entlasten (zu weiteren Informationen über dieses Projekt s. [4]). Im Unterschied zu den im Projekt mit dem Unternehmen Magna Steyr dargestellten Aspekten von Grid Computing, wird in diesem Szenario nicht eine Berechnung sondern eine ganze Arbeitssitzung eines Entwicklers auf die Ressource geleitet, die der Aufgabe am besten gewachsen ist. Dabei spielt nicht nur der momentane Auslastungszustand der Systeme sondern auch Umgebungsfaktoren eine Rolle. Zu Beginn einer Arbeitssitzung muss der Entwickler über CVS eine initiale Umgebung auf dem ihm zugewiesenen System aufbauen. Damit dieser Neuaufbau nicht unnötig wiederholt wird, erhält ein Entwickler nur ein neues System zugewiesen, wenn entweder die Kapazität des Systems es erzwingt, oder die auf dem alten System vorhandenen Daten weit aus der Synchronisation gelaufen sind.

Dieser Fall zeigt sehr deutlich, wie Geschäftsprozesse und Service Level Agreements direkten Einfluss auf die Architektur eines Grid-Systems nehmen.

Sowohl das Referenzbeispiel Magna Steyr als auch Siemens Mobile machten Änderungen in der Anwendungsschicht notwendig. Folgerichtig liegt ein Schwerpunkt der Bemühungen im Bereich

Grid Computing in der Zusammenarbeit mit den unabhängigen Software Herstellern (Independent Software Vendor, ISV).

In den Grid Projekten der IBM wurde nicht nur die Ressource "Rechenleistung" für den einfachen Zugriff bereitgestellt, sondern auch das Thema "Collaborative Engineering" sehr stark fokussiert; in diesem Anwendungsbereich geht es schwerpunktmäßig um die Bereitstellung von Daten für Teams von Ingenieuren im Umfeld der Produktentwicklung. So hat IBM, insbesondere das IBM Labor der IBM Deutschland Entwicklung GmbH in Böblingen, einen Prototyp entwickelt, mit dem Simulationsdaten-Management Systeme der Firma MSC.Software GmbH (www.mscsoftware.com) gekoppelt werden können. Dies erlaubt es Ingenieuren, in allen Instanzen dieser Simulationssteuerungs-Plattform zu arbeiten und auf Projektdaten aller verbundenen Instanzen nahtlos zuzugreifen. Hierdurch wird eine eng verzahnte Zusammenarbeit möglich, die mit der gewonnenen Flexibilität hilft, Kosten zu sparen. Diese gemeinschaftliche Entwicklung von MSC.Software, Audi und IBM wurde in dem EU FP6 Projekt „Simdat" (www.simdat.org) präsentiert und wird in diesem Rahmen weiterentwickelt [5].

Der Charme der Lösung beruht auf dem Einsatz von Standard-Anwendungen, die nicht dediziert für den Einsatz in einem Grid-Kontext erstellt werden mussten. Aus dem Software-Portfolio der IBM konnte für diesen Prototyp die DB2 Datenbank und MQSeries (s. [6,7] für weitere Informationen zu diesen Systemen) als Ergänzung für eine besonders verlässliche Replikation genutzt werden. Die Replikations-Engine ist die verlässliche Grundlage, über die das Simulationsmanagement System erweitert und engmaschig verknüpft werden konnte. Das Zusammenfügen dieser Komponenten ermöglichte es, die vielen Standorte zu einem großen virtuellen Standort zusammenzufassen. Diese neue virtuelle Organisation "Simulation Data and Process Management" (SDM) reagiert robust auf Veränderungen in der Infrastruktur und skaliert von einer kleinen Organisationseinheit zu einer weltweit vernetzten Infrastruktur.

Die Vorteile einer solchen Lösung bestehen in der Flexibilität, nun mehrere Projekte gleichzeitig behandeln zu können, da verschiedene Standorte zu jeder Zeit Zugriff auf alle Daten erhalten. Die streng sequentielle Abarbeitungsreihenfolge der Entwicklungs- und Design-Schritte wurde aufgehoben. Des Weiteren

1.2 IBM Grid Implementierungen

wurde die Wiederverwendung von im Unternehmen bewährten Komponenten durch dieses System erleichtert und gefördert.

Die demonstrierte Flexibilität konnte von Audi direkt mit einem finanziellen Einsparungspotential benannt werde. Neben dieser direkt zugänglichen Größe zählen hier Verbesserungen der Qualität im Design durch höhere Effizienz und schnellere Reaktionszeiten in dem Entwicklungsprozess. Hierzu ein Zitat des Leiters "Functional Design Crash, Occupant Safety" der AUDI AG, Dr.-Ing. U. Widmann: "Collaborating with our engineering team at SEAT will become more and more important due to the common activities inside the Audi brand group. The new technology we are developing with IBM and MSC.Software will help us collaborate more effectively and build a real global engineering collaboration, reducing travel and taking both time and cost out of the product development process".

Neben dem industriellen Sektor ist vor allem der Financial Services-Sektor (FSS) ein Bereich, der intensiv die Nutzung von Grid-Technologien erprobt beziehungsweise diese Technologie schon heute nutzt.

Im Financial Services-Sektor dominieren Monte Carlo-Simulationen zur Berechnung von Risiken wie z.B. „Value at Risk" oder wahrscheinlichen Kursverläufen bei Derivaten oder strukturierten Produkten. Monte Carlo-Simulationen sind als unabhängige Ereignisse zu sehen und so mit sehr geeignet für Verteilungsprozesse. Hier ist die Grid-Architektur eine geeignete Lösung, sehr exakte und aufwändige Berechnungen auf große Computer Verbünde zu übertragen. Im Gegensatz zu den Computer Aided Engineering (CAE) (s. [8] für eine Begriffsbestimmung) Berechnungen in der Automobil- oder Flugzeug- Industrie wird hier auch auf die Forderung nach sehr geringen Latenzzeiten in der Kopplung der Systeme verzichtet. So ist hier ein Grid-System basierend auf Cluster-Systemen und angegliederten vernetzten Workstations denkbar und auch schon bei Kunden verwirklicht worden. In diesem Sektor ist neben dem Life Sciences-Bereich deshalb eine starke Präsenz der Grid Middleware-Anbieter, die auch Workstation-Grids propagieren, zu sehen.

Ein sehr deutliches Beispiel für ein High Performance Grid Cluster ist das 1500-CPU-Cluster bei der Bayrischen HypoVereinsbank.

Die Herausforderung bei der HypoVereinsbank bestand im Bedarf an einer Lösung, mit Hilfe der massiv-parallel laufende Berechnungen schnellstmöglich durchgeführt werden können. Da die Programme im eigenen Haus entwickelt worden sind, konnte man die Programme so optimieren, dass sie am besten in einer eng verknüpften Grid-Infrastruktur laufen konnten und die besten Leistungen zeigten. Die Infrastruktur besteht aus 1.500 Prozessoren, die in der Lage waren, innerhalb von fünf Minuten betriebsfähig zu sein. Dieses System ist hochgradig flexibel, organisiert sich selbst und bringt die höchsten Leistungen. Das gesamte System agiert fehlertolerant durch den Einsatz von automatischen Re-Konfigurationen.

Das führte dazu, dass die Geschäftslogik und die Bereitstellung und Änderungen des gesamten Systems von einem zentralen Punkt aus gesteuert und geändert werden kann.

Durch die Verwendung von Grid-Technologie konnte im FSS eine neue Stufe der Leistungsfähigkeit bei Berechnungen in den Projekten realisiert werden. Waren bisher Grid-Systeme vor allem für Protagonisten der Grid-Innovation wie Ingenieure und die Forschung in öffentlichen Institutionen vorbehalten, die bei dem Entwicklungsprozess die Anfälligkeiten von sehr komplexen Installationen wie Linux Clustern mit mehr als 1000 CPUs in Kauf genommen haben, so musste im Financial Services-Bereich die Hochleistungsfähigkeit der Grid-Systeme mit der Verlässlichkeit und Verfügbarkeit der traditionellen Bankanwendungen zusammengeführt werden.

Als Beispiel dafür dient die Installation einer Grid-Infrastruktur bei dem Bankhaus Sal. Oppenheim (s. [9] für weitere Informationen zu diesem Projekt). Hier wurde das Management des Linux Clusters mit Hilfe der Einbindung einer Tivoli Enterprise-Konsole (IBM Tivoli TEC [10]) und deren Überwachungsprogrammen bewerkstelligt. Des Weiteren war die Einbindung in das bestehende Backup-System der Bank gefordert, sowie ein hochverfügbares Cluster Management und der Möglichkeit des Betreibens eines Patch-Managements mit Versionskontrollen des einzusetzenden Betriebssystems Linux (SuSE Linux Enterprise Server 8.1). Das führte zu einem hoch komplexen System, das auf die Kundenwünsche maßgeschneidert wurde.

Der Nutzen in das Investment des aufwendigen und flexiblen Systemmanagements ist die Möglichkeit, das System dynamisch um neue Systeme erweitern zu können. So wird es mit diesem

Cluster möglich sein, dynamische Kapazitäten, wie sie z.B. von IBM angeboten werden, in das Cluster zu integrieren, da mit minimalem Aufwand das System um entfernte (dezentrale) Knoten (und damit mit zusätzlicher Rechenleistung) erweitert werden kann.

In einer gewachsenen Infrastruktur bei Kunden war es traditionell so, dass neue Software-Produkte auf neuer Hardware installiert worden sind. So wächst schnell der zu betreibende IT-Bereich und verliert an Flexibilität. Die bei Sal. Oppenheim betriebene Softwarelösung für Risiko- und Preis-Berechnungen war nicht mehr in der Lage, Spitzenbelastungen zu handhaben und den Benutzern das Ergebnis in einer angemessenen Zeit zurückzuliefern. Eine einfache Aufstockung der Hardware-Ressourcen war aus der Preis-Leistungs-Betrachtung nicht wirtschaftlich, so dass neue Lösungen gefordert waren. Sal. Oppenheim entschied sich für ein hoch leistungsfähiges Grid-Cluster mit der neuesten Software-Lösung Symphony des ISV Platform Computing und einer IBM Cluster-Technologie.

Diese Lösung war dann in der Lage, dynamisch mit den Bedürfnissen der Bank zu wachsen und die kritischen Geschäftsanwendungen wurden durch nun garantierte Service-Levels zur Verfügung gestellt und eingehalten.

1.3 Schwerpunkt des Grid Computing bei IBM

Wie die kleine Auswahl der IBM Grid-Referenzen im deutschsprachigen Raum aufzeigt, liegt die Herausforderung für die industrielle Nutzung der Grid-Technologien in der Verfügbarkeit der ISV-Anwendung für Grid-Systeme. Im industriellen Bereich sind die Programm-Laufzeiten der Grid Jobs eher in Stunden als in Minuten zu messen. Darüber hinaus liegt die Herausforderung oft in den Latenzzeiten des verwendeten Netzwerks zur Kommunikation sowie in den großen Ein- und Ausgabe-Dateien für die Programme. Im FSS werden, gerade bei analytischen Funktionen die oft auf Monte Carlo-Simulationen beruhen, Jobs mit kurzen Laufzeiten und geringen Abhängigkeiten untereinander (trivial parallelisierbare Anwendungen) und geringen Speicheranforderungen erstellt.

Für diese Funktionen ist das traditionelle Verteilen von Jobs nicht mehr geeignet. Wenn für die Aufgaben-Erstellung eines Jobs nur ein Zeitfenster von wenigen bis 100 Millisekunden zur Verfügung steht, kann dies nur noch durch eine direkte Kommunikation der Anwendungen zu den benutzten Knoten des Grid-Systems geschehen. Hier kommen folgerichtig nicht mehr standardisierte Aufrufe der Anwendung hinzu. Dies ist momentan eine problematische Entwicklung, da es zeigt, dass die Anwendungen sich stärker an die Grid-Middleware binden müssen, als es den ISV recht sein kann.

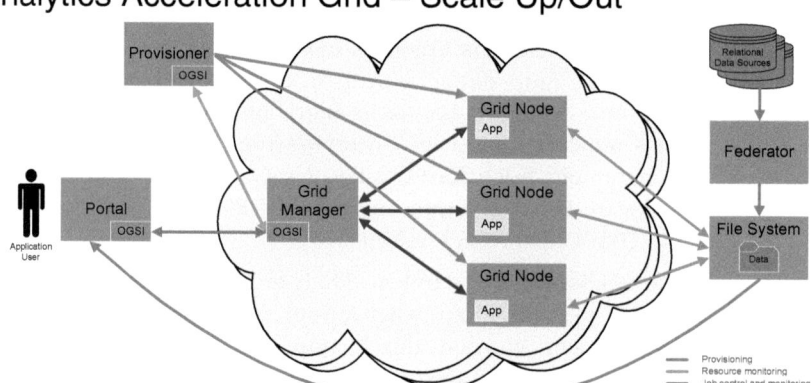

Abb. 2: IBM Analytics Acceleration Grid [11] mit OGSI Steuerungselementen

Große Unternehmen sind darauf angewiesen, dass sie sich durch ihre Architekturrichtlinien ein Höchstmaß an Flexibilität, aber auch den geringsten Integrationsaufwand sichern müssen, um die interne IT als Wettbewerbsvorteil einsetzen zu können.

Wenn nun die Grid-Steuerungen eine enge Verbindung mit den Anwendungen eingehen, kann es passieren, dass das Unternehmen mehrere verschiedene Grid-Steuerungen einsetzen muss.

Hier kommt nun der aktuelle Standard Web Service-Resource Framework (WSRF, [12]) der Globus Alliance (www.globus.org) ins Spiel.

1.3 Schwerpunkt des Grid Computing bei IBM

IBM hat frühzeitig ein internes Projekt gestartet, um alle relevanten Komponenten im Bereich Grid Computing, Beschaffung von Systemen („System Provisioning") und Virtualisierung miteinander zu kombinieren.

In dem von IBM erstellten Angebot "Analytics Acceleration Grid" ist ein Kunde in der Lage, in seinem Unternehmen bestehende Grid-Infrastrukturen unter einer zentralen Verwaltung zu betreiben, die in der Lage ist, auch Lastverteilungsherausforderungen über die Grenzen eines Grid hinweg zu optimieren. Das System sieht vor, dass es eine externe Instanz in diesem System gibt („Versorger" oder „Provisioner" genannt), der Auskunft über die Auslastung aller in den einzelnen Grids eingesetzten Ressourcen bekommt. Werden in einem Grid mehr CPU-Ressourcen benötigt, werden diese durch die Grid-Middleware von einem Grid, das nicht ausgelastet ist, in ein anderes Grid, das einen Ressourcen-Engpass hat, verschoben. Sollte sich zu einem späteren Zeitpunkt wieder die Auslastung im gesamten Grid verschieben, wird dies vom System erkannt und dynamische Adaptionen vorgenommen; dies alles wird durch Elemente bewerkstelligt die dem Grid-Standard konform sind und über diese Wege miteinander kommunizieren.

Große Unternehmen können so in ihrem Rechenzentrumsbetrieb die installierten Ressourcen dynamisch den Aufgaben zuordnen. Hierbei ganze Systeme von Grund auf neu aufzubauen (Network Operating System, Grid Middleware und Anwendung) ist die komplexeste Herausforderung. Es kann aber sinnvoll sein, neue Anwendungen und Grid-Agenten auf die jeweiligen Systeme zu übertragen und zu konfigurieren. Von außen betrachtet wird das "intelligente" Grid als eine von mehreren Ressourcen noch einmal vom übergeordneten Systemmanagement umfasst und verwaltet.

1 Grid Computing für virtualisierte Infrastrukturen

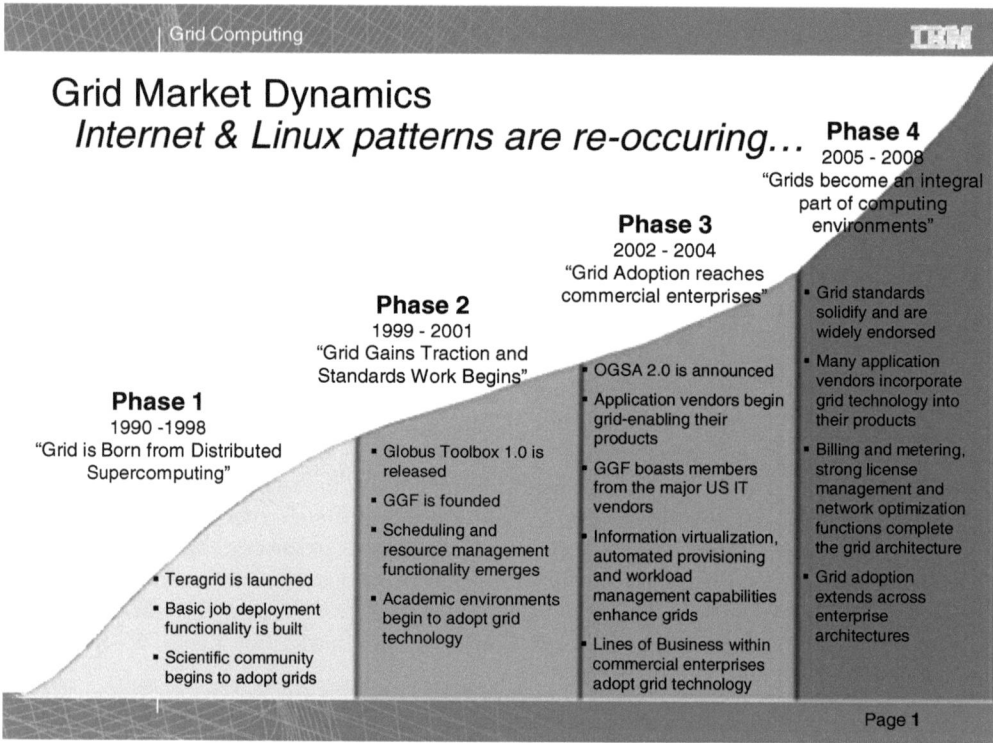

Abb. 3: *Phasen des Grid Computing*

Bei all diesen Vorhaben ist die Zusammenarbeit mit den ISVs eine notwendige Vorraussetzung. Nicht nur das Thema Lizenzkosten muss neu überdacht werden, auch das Design einer Anwendung muss flexibel genug sein, um diesen dynamischen Paradigmen genügen zu können.

In Abb. 3 befinden wir uns folgerichtig in der Phase des Aufbaus eines erfolgreichen Systems mit ISV und System-Integratoren, deren Geschäftsfeld die Erbringung von IT-Dienstleistungen und Integration von Prozessen, Anwendungen und IT-Infrastrukturen ist.

1.4 Grid-Infrastrukturen der Zukunft

Alle bisher betrachteten Grid-Infrastrukturen basierten auf der Herausforderung, innerhalb eines Unternehmens eine Grid-Lösung aufzubauen. Diese Grid-Lösungen kann man grob in drei Klassen unterteilen:

1.4 Grid-Infrastrukturen der Zukunft

1. Die Grids, die auf einer Cluster-Infrastruktur aufbauen und nur homogene Komponenten verwenden, wie beispielsweise x-mal den Prozessor der Marke a.

2. Jene Grids, die auf der Verwendung von heterogenen Ressourcen aufbauen, aber nicht den Rahmen einer Abteilung oder eines IT-Shops überschreiten; eine Abteilung kann hier über den Einsatz der Ressourcen und verwendeten Software-Lösungen entscheiden.

3. Unternehmensweite Grids, wie im Beispiel des Analytics Acceleration Grid, das es zur Aufgabe hat, verschiedene Grid-Lösungsinseln eines Unternehmens in einen größeren Verbund und Kontext zu stellen (s. Abb. 2).

Wenn man allerdings den Rahmen und Geltungsbereichs des eigenen Unternehmens verlassen will, muss man sich mit anderen Aspekten einer Grid Infrastruktur auseinandersetzen, über die man sich vorher nur sekundär Gedanken machen musste. An dieser Stelle sollen nur zwei Stichworte - Sicherheit und Abrechnungs-Modelle - aufgeführt werden.

Wenn wir nun diesen Grid-Ansatz weiter betrachten, der nicht mehr an den Grenzen des eigenen Unternehmens halt macht, so kann man sich folgenden Aufbau vorstellen:

Jede bisher betrachtete Grid-Implementierung machte es sich zu Nutzen, dass Regeln in der Benutzung der Ressourcen gefunden werden mussten. Es wurde ein System definiert, wie die Ressourcen untereinander kommunizieren konnten und was der gemeinsame Zweck und Nutzen der Infrastruktur war. Es gibt immer eine zentrale Instanz, die Verwaltung und Überwachungsfunktionen wahrgenommen hat, sowie Knoten von Computern oder Cluster Systemen, die die anfallende Rechenlast bereitstellten. Nimmt man dies auf, kann man sich folgende flexible und skalierbare Architektur vorstellen:

Es gibt eine zentrale Komponente, die verschiedene Aufgaben übernimmt, die in einem globalen, weit vernetzten Grid nötig sind.

Ein Beispiel einer zentralen Rolle ist die Public-Key-Infrastructure (s. z.B. [13]), die die Rolle hat, die Zertifikate zu erstellen, die nötig sind, um einen rechtmäßigen Zugriff auf Ressourcen aller Art zu steuern. Diese Certificate Authority (s. z.B. [13]) gibt es nur einmal und wird von allen Parteien dieser Infrastruktur als die maßgebliche angesehen. Andere wichtigen Funktionen sind darüber hinaus noch Namensräume und auch die Verwaltung der in diesem Grid angeschlossenen Mitspieler, die Ressourcen beisteuern.

Jede neue Komponente, die in dieses Grid-System integriert wird, muss sich als erstes an der zentralen Instanz anmelden, bekommt so ihre Identität, ist innerhalb des Verbundes bekannt gegeben und kann Ressourcen aus der eigenen Umgebung benutzen, aber auch entfernte Ressourcen anderer Komponenten benutzen. Hierfür ist die „Grid-Zentrale" die Clearing-Stelle, um bei Ressourcen-Engpässen einer Komponente, die nicht in der Lage ist, einen Job lokal abzuarbeiten, mit Ressourcen einer entfernten Grid-Komponente zu verbinden, welche bei der Bearbeitung der Aufgabe mit verwendet werden können.

Ein wichtiger Punkt bei dieser Betrachtung ist, dass es eine Grid-Infrastruktur für die Datenhaltung gibt, sodass die Daten, die ein Benutzer auf „seiner" Ressource lokal gespeichert hat, nicht unbedingt explizit zu entfernten Ressourcen übertragen werden müssen, die bei der Bearbeitung des Jobs genutzt werden. Eine Lösung wie das General Parallel File System (GPFS, [14]) von IBM hilft hier, die Infrastruktur einfach zu halten. Damit ist es möglich, dass die Daten bei den Benutzern bleiben. Die Verwaltung des GPFS, was auch den Namensraum der Daten angeht, wird zentral von der Grid-Zentrale verwaltet.

Die Grid-Zentrale bekommt von den Grid-Komponenten in regelmäßigen Abständen den Auslastungsgrad zugeschickt, sodass die Grid- Zentrale alle ankommenden Anfragen nach Bearbeitung von Jobs beantworten kann. Über eine Routine, die den „Fair Share" berechnet, kann eine gleichmäßige Auslastung der gesamten Infrastruktur gewährleistet werden.

Um auch die Auslastung transparent für alle Teilnehmer des Grids zu machen, erstellt die Grid-Zentrale eine Statistik, die auch weiter verdichtet werden kann, um ein Abrechnungsmodell zu implementieren. So werden die Ressourcen in einem „globalen" Grid dieser Art transparent.

Sollte an einer Stelle die Ressourcen-Kapazität nicht ausreichen, kann man durch Erweiterung des modularen Systems jeder Zeit mehr Hauptspeicher oder Datenspeicher oder Netzwerkadapter o.ä. einbauen oder auch durch neuere Komponenten austauschen.

IBM entwickelt aufgrund dieser Anforderungen eine Architektur, in der die einzelnen Elemente der Grid-Komponenten mit einer geprüften Verträglichkeit kombiniert werden können.

Vordefinierte, angepasste Softwarestände können dann von einer Grid-Zentrale bezogen werden und die isolierte Installation einer Grid-Komponente in das gesamte System überführt werden.

1.5 Grid und Virtualisierung

Grid-Technologie bietet die Konzepte und Mittel, Anwendungen zu virtualisieren. Die logische Weiterentwicklung davon ist das Virtualisieren der gesamten IT-Infrastruktur.

Das bedeutet nicht, dass nur noch eine Standardinfrastruktur existiert, die allen Anforderungen dynamisch gerecht wird. Vielmehr müssen die heute sehr komplexen und leistungsfähigen Anlagen „lernen", sich den Anforderungen der Anwendungen anzupassen. Dies kann heißen: Ressourcen von untätigen, geladenen Anwendungen hin zu kritischen Performance-Engpässen umleiten, automatisch mehrstufige Systeme (Web-Server, Anwendungs-Server und Datenbank-Server) in ihrem gesamten Service-Level für die Anwender überwachen und Empfehlungen aussprechen.

All diese Aufgaben hat IBM in die Elemente einer Initiative namens "Virtualization Engine" [15] integriert.

In den kommenden Jahren wird das Thema Virtualisierung heterogener IT-Ressourcen bei strikter Standardkonformität eine zentrale Aufgabe der IBM sein.

1.6 Literaturverzeichnis

[1] S. Zhou. LSF: Load-sharing in large-scale heterogeneous distributed systems. In: Proceedings of the Workshop on Cluster Computing, Orlando/FL, April 1992.

[2] P.T. Bulhoes, C. Byun, R. Castrapel, O. Hassaine, N1 Grid Engine Features and Capabilities, Beitrag bei der SUPerG 2004, Phoenix/AZ, Mai 2004 (http://www.sun.com/products-n-solutions/edu/whitepapers/pdf/N1GridEngine6.pdf (Abruf am 11.10.2005)

[3] http://www-5.ibm.com/de/pressroom/presseinfos/2004/040331_1.html (Abruf am 7.10.2005)

[4] http://www-5.ibm.com/de/pressroom/presseinfos/2004/040917_2.html (Abruf am 7.10.2005)

[5] J. Reicheneder et al., Definition of SAMD (SIMDAT Automotive Demonstrator) Reference Model and Use Case, http://www.scai.fraunhofer.de/fileamin/images/nuso/SIMDAT/SIMDAT_D.11.1.1_public.pdf (Abruf am 11.10.2005)

[6] http://www.ibm.com/software/de/db2/ (Abruf am 12.10.2005)

[7] http://www.ibm.com/software/info/ecatalog/de_DE/products/G106020C87422R91.html?&S_TACT=none&S_CMP=none (Abruf am 7.10.2005)

[8] G. Spur, F.-L. Krause, Das virtuelle Produkt – Management der CAD-Technik, Hanser-Verlag, 1997.

[9] http://www-1.ibm.com/grid/grid_press/pr_1216.shtml (Abruf am 7.10.2005)

[10] http://www-306.ibm.com/software/tivoli/products/enterprise-console/ (Abruf am 7.10.2005)

[11] R. Vrablik, R. Stryniewicz, C. Reech, D. Spexet, Analytics Acceleration Grid Environment – How grid systems enable those applications, http://www.ibm.com/developerworks/grid/library/gr-aage1/ (Abruf am 11.10.2005)

[12] K. Czajkowski, D. Ferguson, I. Foster et al., The WS-Resource Framework 1.0, http://www-128.ibm.com/developerworks/webservices/library/specification/ws-resource/ws-wsrfpaper.html (Abruf am 11.10.2005)

[13] C. Eckert, IT-Sicherheit – Konzepte – Verfahren – Protokolle, 3. A., Oldenbourg, 2004

[14] F. Schmuck, R. Haskin, GPFS: A Shared-Disk File System for Large Computing Clusters, In: Proc. of the 1st Conf. on File and Storage Technologies (FAST), 28./29.1.02, Monterey/CA

[15] http://www-03.ibm.com/servers/eserver/about/virtualization/access.html (Abruf am 7.10.2005)

2 Service Grids – von der Vision zur Realität

A. Geiger

Die Grid-Technologie ist heute am sichtbarsten in der Forschung. Eine Vielzahl von Projekten versucht das Thema weiter zu treiben und Standards zu etablieren. Ziel ist eine Infrastruktur, bei der es für jede Person oder Organisation möglich ist, zu jeder Zeit und an jedem Ort auf die Ressourcen zuzugreifen, die gerade benötigt werden.

Der Produktionseinsatz von Grids beschränkt sich heute jedoch noch auf die Verbesserung der Flexibilität und Auslastung von Daten- und Rechenzentren. Der Endnutzer spürt dies an besseren Qualitätseigenschaften und attraktiveren Preisen für Services. Von der Grid-Technologie selbst merkt er so gut wie nichts.

Service-Grids hingegen werden unmittelbar und einschneidend die Geschäftsmodelle für ITC-Services verändern.

In diesem Kapitel soll diese Zukunftsvision skizziert und der Stand der Entwicklung aus der Sicht von Technik und Management beschrieben werden.

2.1 Einführung

Die Abbildung von Geschäftsprozessen mit Mitteln und Methoden der Informations- und Telekommunikationstechnologie (ITC) wird heute unter dem Themenkomplex eBusiness zusammengefasst. Stand das `e´ in der Anfangszeit noch für 'electronic', so wurde im Laufe der Zeit schnell klar, dass die Entwicklung nicht bei der Abbildung bestehender Geschäftsprozesse stehen bleiben würde, sondern dass die eingesetzten Werkzeuge der ITC die Möglichkeit bieten, diese Prozesse funktionell deutlich weiterzuentwickeln. Heute versteht man konsequenterweise unter eBusiness deshalb 'enhanced Business'.

Je nach Einsatzbereich bildeten sich im Laufe der Zeit Spezialisierungsrichtungen des eBusiness heraus, beispielsweise eCommerce für den ITC-gestützten Handel, eEngineering für den Bereich der Entwicklung technischer Produkte und eScience für die Wissenschaft.

In allen Bereichen steht das `e´ für ein Mehr an Funktionalität. Standen beispielsweise in den Ingenieurwissenschaften die so genannten CAx-Technologien für die computergestützte Lösung technischer Probleme, so ist eEngineering weit mehr, nämlich ein Geschäftsprozess der neben den technischen auch die kaufmännischen, administrativen und sonstigen involvierten Disziplinen umfasst.

Ähnlich sieht es in der Wissenschaft aus. Vor geraumer Zeit startete mit Computational Science der Siegeszug der Simulation als dritte Methodik der wissenschaftlichen Erkenntnisgewinnung neben Experiment und Theorie. Konsequenterweise fasst eScience diese drei Basismethoden wieder zu einem Geschäftsmodell zusammen, dessen Möglichkeiten weit über die der drei Einzelmethodiken hinausreichen und das auch hier nicht mehr nur die technisch-wissenschaftlichen Aspekte umfasst.

Da sich eScience und eEngineering als Methodiken und Basis für Geschäftsmodelle auf Werkzeuge der ITC stützt, brauchen sie eine passende technische Infrastruktur. Natürlich bilden Rechner als Grundelemente der Verarbeitung und das Internet als Medium für den Transport von Daten dafür die Basis. Als Zwischenschicht und somit als sichtbare Schnittstelle der technischen Infrastruktur wird jedoch eine Umgebung benötigt, die es erlaubt, beliebige Ressourcen als Teil von wissenschaftlichen Geschäftsprozessen zu nutzen. Die dafür notwendige Basistechnologie stellt das Grid bereit.

2.2 Vision von eScience und eEngineering im Service-Grid

Um Forschung und Entwicklung neue Horizonte zu öffnen, ist es notwendig, ITC-Leistungen in sehr flexibler und dynamischer

2.2 Vision von eScience und eEngineering im Service-Grid

Form mit höchster Zuverlässigkeit zur Verfügung zu stellen. Leistungen verschiedener Lieferanten müssen des Weiteren beliebig und ohne Zusatzaufwand zu einem Gesamtsystem verknüpfbar sein, das das aktuelle Vorhaben in all seinen Teilaspekten ITC-mässig abbildet. Im Gegensatz dazu bedeutet der Bezug von Leistungen heute die Durchführung von Ausschreibungsverfahren und den Abschluss von Verträgen mit einer geringen Flexibilität, Quantität und Qualität an die aktuellen Bedürfnisse anzupassen.

Diese Flexibilität in der dynamischen Bereitstellung untereinander kompatibler ITC-Leistungen lässt sich nur durch den Einsatz autonomer, also sich selbständig überwachender und rekonfigurierender Systeme erreichen, in denen dynamisch und bedarfsgesteuert Ressourcen allokiert und freigegeben werden. Auch die Einhaltung der notwendigen Qualitätsmerkmale (Service-Levels) wird in einem autonomen System durch den automatisierten Ausbau bzw. Austausch von Komponenten garantiert. Gerade in komplexen und wenig standardisierten Systemen hilft es wenig, wie heute die Einhaltung von Qualitätsparametern zu berichten und anschliessend in einen langwierigen Verbesserungsprozess zu gehen. Stattdessen muss ein autonomes System von seiner Konzeption her die Einhaltung garantieren und bei SLA-Verletzungen die Komponenten einfach austauschen.

Die Vision wäre aber nicht vollständig, beschränkte man sich allein auf funktionale Aspekte. Wirklicher Fortschritt wird auch am effizienteren und effektiveren Einsatz von Ressourcen gemessen. Gerade bei externem Bezug von Leistungen ist es wichtig, auch in autonomen Systemen den Wettbewerb unter den Anbietern aufrechtzuerhalten oder gar zu forcieren. Hierzu benötigen autonome Systeme eine Broker-Komponente, die bei der Allokation von Ressourcen die qualitativen und wirtschaftlichen Parameter optimiert und autonom den optimalen Lieferanten auswählt. Der Aspekt der gemeinsamen Nutzung von Ressourcen (Shared Services) wirft allerdings zusätzliche Fragestellungen auf.

In einer Umgebung, wie sie soeben beschrieben wurde, werden ITC-Leistungen nicht mehr durch den Kauf von Komponenten und Services, sondern extern aus dem Netz bezogen. Der für die Lösung einer Aufgabe zusammengestellte Gesamtkomplex überspannt also in der Regel mehrere Organisationen (Lieferanten und Projektpartner). Man spricht hier auch von einer virtuellen

Organisation. Da sich in einem solchen Gesamtszenario u.U. nicht nur vertraute (oder vertrauenswürdige) Kooperationspartner finden, ist es essentiell, dass die Schnittstellen zwischen den Komponenten in hohem Masse die heute üblichen Sicherheitsstandards reflektieren, insbesondere was die Vertraulichkeit (Confidentiality) betrifft.

Die Diskussion über die Vor- und Nachteile eines klassischen Outsourcing von ITC-Services wird sich in Zukunft also in ganz anderer Form stellen als heute. Für Basisleistungen wird die Diskussion auf jeden Fall aufgrund der technischen Gegebenheiten obsolet. Es geht in Zukunft nur noch um die Frage, ob die Integration und Betreuung der Geschäftsprozesse ausgelagert oder selbst erledigt wird.

2.3 Die Basis: Grid-Technologie

2.3.1 Internet – Web – Grid

Die Begriffe Internet, Web und Grid werden, insbesondere im deutschen Sprachraum, von ihrer Bedeutung her oft nicht sauber voneinander abgegrenzt. Dies führt leicht zu Erklärungsnöten wenn es darum geht, die Bedeutung der Grid-Technologie herauszuarbeiten. Die Gründe für diesen Zustand mögen zum einen darin liegen, dass sich bei neuen Technologien exakte Definitionen und Abgrenzungen erst im Laufe der Zeit entwickeln, zum anderen aber auch darin, dass die Nutzung des Internet, das seit ca. 1969 existiert, in Deutschland erst durch die Web-Technologie nach 1990 auf breiter Front populär wurde. Umso wichtiger ist es aber an dieser Stelle, die Begriffe klar voneinander abzugrenzen:

- **Internet**
  Das Internet ist eine Infrastruktur bestehend aus Netzverbindungen und standardisierten Protokollen, um Daten zwischen beliebigen Geräten über beliebige Distanzen auszutauschen.
- **Web**
  Das World Wide Web (WWW) erlaubt den beliebigen und weltweiten Zugriff auf Informationen. Es bedient sich des Internet als Transportmedium.
- **Grid**
  Das Grid erlaubt den beliebigen und weltweiten Zugriff auf alle Arten von Ressourcen. Es bedient sich des Internet

als Transportmedium. Das Grid ist somit eine Generalisierung des Web, indem es die Klasse zugreifbarer Resourcen auf Rechenleistung, Daten, Instrumente, Sensoren, Anlagen, Services etc. ausdehnt. Konsequenterweise ist die Grid-Technologie damit die Basistechnologie für organisationsübergreifende Geschäftsprozesse und die gemeinsame und damit wirtschaftlichere Nutzung von Ressourcen.

2.3.2 Konvergenz von IT und TC

Das Grid war in seinen Anfängen primär durch den Zugriff auf Rechenleistung geprägt. Zur selben Zeit entwickelte sich die Web-Technologie durch die Einführung von Portalen zur Nutzung verschiedenartigster Ressourcen im Netz ebenfalls weiter. Da die frühe Grid-Community fest im technisch-wissenschaftlichen Rechnen verwurzelt war, verliefen diese beiden Entwicklungen nahezu unabhängig voneinander.

Durch die immer stärker werdende Einbeziehung der Industrie und somit einem Aufkommen kommerzieller Interessen wurde aber schnell klar, dass die Entwicklung wesentlich beschleunigt werden kann, wenn beide Communities voneinander lernen und bereits vorhandene Lösungen jeweils übernommen werden. Dies führte schliesslich zu der oben angegebenen Definition des Grid als Generalisierung des Web.

2.3.3 Ausprägungen von Grids

- **Compute-Grids**
  Ziel eines Compute-Grids ist in der Regel die verteilte Bereitstellung einer Rechenleistung oder einer Rechenkapazität, die dem Anwender nicht in seiner eigenen Umgebung zur Verfügung steht. Dabei kann es um die Nutzung ansonsten brachliegender Ressourcen in der eigenen Organisation (z.B. Arbeitsplatzrechner außerhalb der üblichen Geschäftszeiten) oder die Lösung von Extremproblemen auf zusammen geschalteten Rechenanlagen der höchsten Leistungskategorie gehen.
- **Data-Grids**
  Aufgabe eines Data-Grids ist die gemeinsame Nutzung und Verarbeitung grosser Datenmengen. Wichtigster Aspekt eines Data-Grid ist die Bildung einer Data-Federation, d.h. einer gemeinsamen, organisations- und lokationsü-

bergreifenden Sicht auf alle Daten, die z.B. zu einem Projekt gehören. Dabei gilt als oberstes Gebot, dass derjenige, der Daten in einer solchen Umgebung bereitstellt, die volle Kontrolle über diese Daten behält. Im Gegensatz zu sogenannten globalen Dateisystemen haben wir es also mit einem dezentral gemanagten System zu tun.

- **Application-Grids**
  Waren die ersten Compute- und Data-Grids noch auf eine Organisation, mindestens aber auf eine Sicherheitsdomäne beschränkt, so wurde mit Application-Grids der erste Schritt hin auf das ehrgeizige Ziel virtueller Organisationen getan. Ziel ist die organisationsübergreifende gemeinsame Nutzung von Ressourcen und damit die verbesserte Auslastung für den Betreiber sowie ein breiteres Angebot für den Nutzer. Fragestellungen, die für solche Application-Grids zu lösen sind, sind der sichere und schnelle Datentransport, Authentisierung und Autorisierung, Single Sign-On sowie der Themenkomplex von Accounting und Abrechnung. Obwohl sie in diesem Sinne noch keineswegs vollständig ist, stellt die aus einem BMBF-Projekt hervorgegangene Softwareumgebung UNICORE (Uniform Access to Computing Resources) die heute in diesem Umfeld funktionell anspruchsvollste Lösung dar.

- **Resource-Grids**
  Ein natürlicher Evolutionsschritt ausgehend vom Application-Grid, ist die Einführung eines Rollenmodells. Ein Resource-Grid differenziert klar zwischen der Rolle des Grid-Nutzers, des Grid-Providers und des Resource-Providers. Dabei greift ein Grid-Nutzer auf die Grid-Infrastruktur des Grid-Providers zu und nutzt die dort von Resource-Providern angebotenen Ressourcen. Obwohl die Funktionalität sich für den Endanwender nicht gravierend von der eines Application-Grid unterscheidet, gibt es konzeptionell einen entscheidenden Unterschied: Application-Grids sind in der Regel vertikal integrierte Systeme, bei denen alle Komponenten individuell integriert werden. Ein Resource-Grid hingegen verlangt die Definition und Offenlegung von Schnittstellen. Jeder Resource-Provider muss wissen, welchen Spezifikationen sein Angebot genügen muss, um in der Grid-Umgebung des Grid-Providers angeboten werden zu können. Sinnigerweise standardisiert man diese Schnittstellen (Virtualisierung der Ressourcen).

2.3 Die Basis: Grid-Technologie

- **Service-Grids**
  Ein Service-Grid kombiniert die Technik des Resource-Grid mit dem Konzept nutzerorientierter Services. Ein Service besteht in der Regel aus einer Vielzahl von Komponenten, von denen jede einzelne von einem anderen Resource-Provider als Utility bereitgestellt werden kann. Beim Service-Grid werden die Resource-Provider nicht mehr direkt gegenüber dem Grid-Nutzer exponiert, in der Regel kennt er sie nicht einmal. Der Grid-Service-Provider ist damit eine generalisierte Form des Grid-Providers, indem er für den gesamten Nutzerservice verantwortlich ist, die Resource-Provider auswählt und diesen gegenüber genauso
  wie gegenüber dem Nutzer abrechnet. Weiter obliegt dem Service-Provider der Aufbau des Gesamtservice aus Einzelressourcen (Choreographie). In der folgenden Abb. 1 soll verdeutlicht werden, dass wir es in einem Service-Grid typischerweise mit einer Vielzahl von Resource-Providern zu tun haben, von denen häufig der Kunde selbst einer ist (mit den Hard- und Software-Ressourcen die sich in seinem eigenen Besitz befinden).

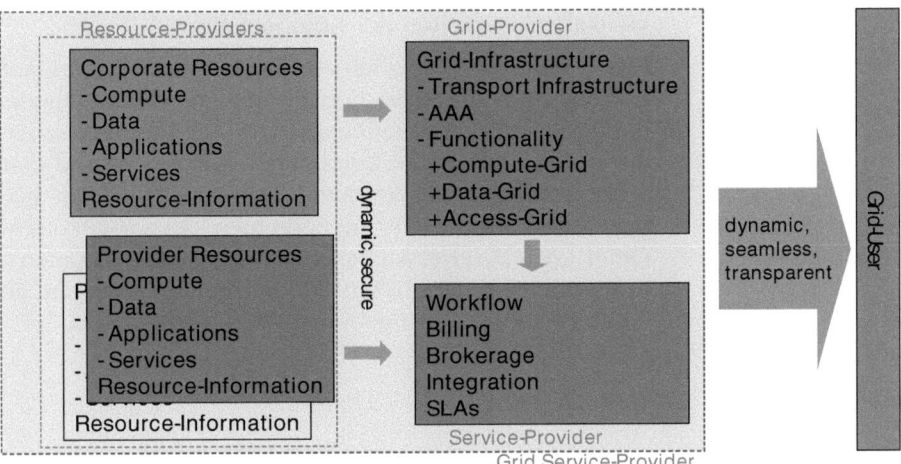

Abb. 1: *Service-Grid [T-Systems]*

Während die Entwicklung bei der Bereitstellung von Rechenleistung, Daten und Anwendungen in einem solchen Service-Grid bereits sehr weit fortgeschritten ist, bleiben die Netze ein Sorgenkind der Entwicklung.

Bisher existiert am Markt keine Netztechnologie, die über weite Leistungsbereiche skalierbar ist, also dynamisch bereitgestellt werden könnte. Auch das Geschäftsmodell der meisten Netzbetreiber basiert auf einer Berechnung der Bereitstellung und nicht der Nutzung. Hier besteht noch gravierender Entwicklungsbedarf sowohl was die Technik, als auch was die wirtschaftlichen Modelle angeht um diese Ressourcen vernünftig in einer Grid-Umgebung als Utility bereitstellen zu können.

2.4 Architektur

Um den Nachweis zu führen, dass der beschrittene Weg wirklich einen Fortschritt bringt, wurde bei den ersten Pilotimplementierungen von Grid-Umgebungen der Fokus auf Funktionalität für den Endnutzer gelegt. Wie bei Pilotimplementierungen in der IT üblich, sollten bei der Integration die Teilkomponenten so gering wie möglich verändert werden. Die so entstandenen Umgebungen liessen sich natürlich nicht ohne Weiteres für andere Anwendungsbereiche übertragen. Wir sprechen deshalb von einer vertikalen Integration der Komponenten.

Um schnell und flexibel eBusiness Umgebungen aufbauen zu können ist es hingegen notwendig, ein Schichtenmodell mit sauber definierten und offengelegten Schnittstellen zu entwickeln um die Komponenten kompatibel zueinander und damit miteinander kombinierbar zu halten. Dieser Aufgabe hat sich das Global Grid Forum GGF angenommen. Herausgekommen ist die Open Grid Service Architecture (OGSA). Ressourcen und Grid-Komponenten, die diesem Modell genügen, lassen sich flexibel zu Problemlösungsumgebungen zusammenbauen.

2.4 Architektur

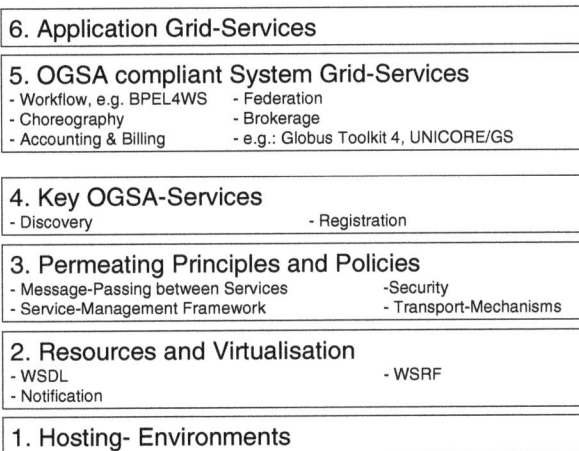

Abb. 2: *OGSA-Schichtenmodell des Global Grid Forum*

Das OGSA-Modell umfasst insgesamt sechs Schichten.

- Virtualisierung
  Der heute für den Produktionsbetrieb wichtigste Aspekt dieses offenen Modells wird dabei auf Schicht zwei behandelt, nämlich die Virtualisierung. Im Gegensatz zu bisherigen verteilten Umgebungen, in denen die technischen Details und der Ort jeder Infrastrukturkomponente gegenüber dem Nutzer exponiert sind, sieht der Nutzer in einer virtualisierten Umgebung nur noch die Eigenschaften einer Leistung, die ihm über eine standardisierte Schnittstelle zur Verfügung gestellt wird.

  Serviceorientierte Architektur

  Werden virtualisierte Services über ein Web-Interface zur Verfügung gestellt, so sprechen wir von einer serviceorientierten Architektur basierend auf Webservices. Die Anbieter solcher Webservices veröffentlichen dazu die Eigenschaften ihrer Services in einer Form, die der Web Services Definition Language (WSDL) entspricht. Nutzer, die einen bestimmten Service suchen, wenden sich dazu an einen Service-Broker, der die Eigenschaften gesuchter und angebotener Services mit Hilfe eines Directory basierend auf

UDDI (Universal Description, Discovery and Integration) miteinander abgleicht. Haben Nachfrage und Angebot zueinander gefunden, greift der Nutzer auf den Service des Anbieters zu.Dies erlaubt die Abbildung von Workflows über System- und Organisationsgrenzen hinweg und die Integration bestehender Anwendungen und Systeme. Die Kommunikation zwischen den Komponenten erfolgt über das XML-basierte Protokoll SOAP (Simple Object Access Protocol) entweder im Stil eines Remote Procedure Call (rpc) oder durch Nachrichtenaustausch. Im Gegensatz zur normalen WWW-Nutzung kommunizieren hier in erster Linie Computer, oder allgemeiner: Ressourcen, miteinander. Dies kann auch Systeme umfassen, die ansonsten im WWW nicht sichtbar sind. In diesem Fall muss allerdings die Sicherheitsarchitektur sorgfältig geplant werden.

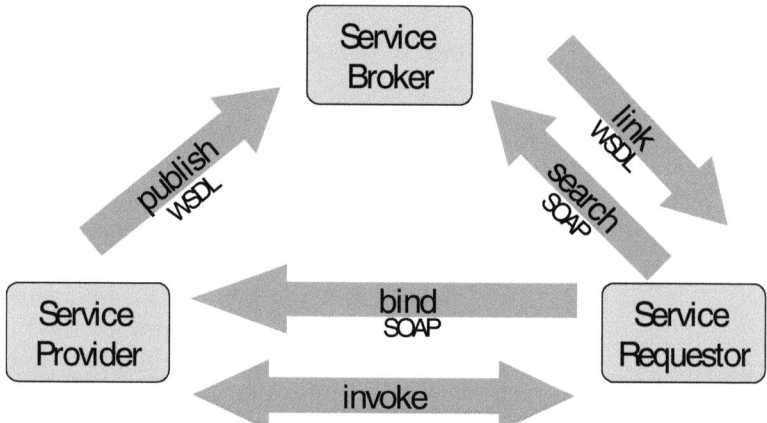

Abb. 3: *Ablauf zum Aufruf eines Web Service*

- Choreographie
  Serviceorientierte Architekturen sind heute bereits in vielen Unternehmen und Organisationen Stand der Technik. Sie weist jedoch immer noch zwei gravierende Defizite auf:
    - Das Auffinden und das Binden der Nachfrage an einen Service sind statisch. Bei Ausfall des Service oder bei Auftauchen eines attraktiveren Angebotes kann nicht unterbrechungsfrei zwischen Anbietern umgeschaltet werden.

o Der Aufbau nicht-persistenter Geschäftsprozesse ist nicht möglich.

Auf den Ebenen 3-5 des OGSA Modells werden deshalb Prinzipien, Basisdienste und Schnittstellen definiert, um komplexe und dynamische Services aus einzelnen Service-Komponenten (Utilities) aufzubauen.

2.5 Auswirkungen

Die Einführung von Service-Grids für Geschäftsprozesse wird heute in erster Linie als technisches Problem betrachtet. Dies ist natürlich, da bei der Einführung der meisten neuen Entwicklungen die gesellschaftlichen und wirtschaftlichen Aspekte erst betrachtet wurden, als alle technischen Probleme gelöst waren. Andererseits ist dies aber auch gefährlich, da die Einführung u.U. dann nicht mehr steuerbar ist.

- Shared Services

  Der zentrale Aspekt bei Service-Grids ist die gemeinsame Nutzung von Ressourcen durch verschiedene Organisationen und Services. Wurde bisher z.B. ein Rechner für die Lösung technisch-wissenschaftlicher Aufgabenstellungen bei <u>einer</u> Organisation gekauft, so kann dieser Rechner als virtualisierte Resource in einem Service-Grid abwechselnd technisch-wissenschaftliche und betriebswirtschaftliche Probleme für <u>verschiedene</u> Organisationen bearbeiten, soweit Hard- und Software dafür geeignet sind.

  Bilden heute Betreiber und Nutzer von Resourcen in vielen Bereichen (z.B. High Performance Computing) noch eine kompakte Einheit, so werden diese Beziehungen in Zukunft aufgelöst. Nicht mehr der ´Besitz´ einer Ressource entscheidet über die Fähigkeit, bestimmte Probleme lösen zu können, sondern das Budget für den Zugriff auf eine entsprechende Ressource am Weltmarkt. Auf jeden Fall rückt eine Service Grid-Umgebung wesentlich mehr den Anwender in den Mittelpunkt der Betrachtungen als den Betreiber der Infrastruktur.

- Sicherheit und Vertrauen

  Diese Arbeitsweise wirft natürlich sofort Sicherheitsfragen auf und zwar sowohl hinsichtlich der Funktionssicherheit (Safety), der Authentisierung und Autorisierung (Security), aber insbesondere die nach der Vertraulichkeit der Information (Confidentiality). Da der Endnutzer in der Regel in einer solchen Umgebung gar nicht mehr weiß, mit wem er die Infrastruktur teilt, muss er sich voll darauf verlassen können, dass die eingesetzten Schnittstellen und Verfahren seine Vertraulichkeitsanforderungen unter allen Umständen gewährleisten. Dazu müssen die Standards offen und kontrollierbar sein.

Service-Grids sind nicht mehr deterministisch was die Nutzung bestimmter Ressourcen betrifft. In vielen sensiblen Bereichen (Sicherheitstechnik, Medizin) erfordern jedoch Unternehmenspolicies oder gesetzliche Regelungen, dass jederzeit und lückenlos über die komplette Verarbeitungskette ein Nachweis erfolgen kann (Audit). Speziell im medizinischen Bereich muss dieser Nachweis bereits erfolgen, bevor die Verarbeitung beginnt (Tracking).

Gerade die dynamische Choreographie erfordert auch einen dynamischen Umgang mit Vertrauen und der Delegation von Vertrauen. Hier kann entweder mit einer expliziten Vertrauensdelegation oder mit Proxies und Vertrauensmetriken gearbeitet werden. In letzterem Fall bedeutet dies aber, dass eine Person oder Organisation die Vertrauensdelegation einer Softwareinstanz überlässt.

Selbst ansonsten einfache Fragestellungen wie die der Datensicherung sind in einem Service-Grid nicht mehr trivial. Unabhängig vom Ort der Verarbeitung muss die Sicherung natürlich an einer Stelle erfolgen, die für den Nutzer zugreifbar ist.

- Lizenzmanagement

    Die organisationsübergreifende Nutzung und Abrechnung von Softwarelizenzen ist ein zwar technisch lösbares Problem, aber leider steht ein solches Modell der nutzungsabhängigen Abrechnung von Software-Nutzung im Konflikt mit den Lizenzmodellen der meisten Softwareanbieter. Hier müssen neue Geschäfts- und Nutzungsmodelle entwickelt werden wie z.B. das in diesem Artikel skizzierte Modell für Application-Service Providing der zweiten Generation (ASP-2).

2.5 Auswirkungen

- Verrechnungsmodelle und Globalisierung

  Um Leistungen in einem Service-Grid abrechnen zu können, muss jede Ressource mit einem klaren Preismodell hinterlegt sein. Dieses kann sehr wohl dynamisch in dem Sinne sein, dass ein Service seinen Preis abhängig von der Nachfrage automatisch ändert. Da wir es mit einem System zu tun haben, bei dem die meisten Organisationen sowohl in der Rolle als Anbieter (Quelle) als auch Nutzer (Senke) von Ressourcen auftreten, wird in einigen Nutzer-Communities auch über Austauschmodelle diskutiert, die letztendlich eine moderne Form des Naturalienhandels darstellen würden. Ein solches Modell könnte jedoch nur in einer geschlossenen Umgebung funktionieren, nicht jedoch in einem weltweiten Markt.

- Kostenreduktion und Strukturelle Veränderungen

  Bisher wurden in der IT hauptsächlich Programmier- und andere Entwicklungsaufgaben in Billiglohnländer verlagert (Offshoring), bei Services war dies wegen des starken Personen- und Prozessbezugs nicht möglich. In Zukunft ist das Offshoring von Servicekomponenten kein grundsätzliches Problem mehr. Nur die Choreographie von Komplettservices und die Abbildung von Geschäftsprozessen müssen noch in Kundennähe geschehen, da hier spezifische Kenntnisse der Nutzerprozesse und eine ständige Interaktion mit dem Kunden und Nutzer notwendig sind.

Im Bereich der ITC-Services kommt also eine ähnliche Entwicklung wie in der Fertigungsindustrie auf uns zu. Prozesse, Produkte und Leistungen werden so stark modularisiert, bis es sich bei den Einzelmodulen größtenteils um austauschbare Standardbausteine handelt. In diesem Falle wird sich der Austausch sogar vollständig automatisieren lassen. Die heute im Bereich der ITC-Services üblichen langfristigen Vertragsbeziehungen machen nur noch für die Integration der Module zu nutzernahen Workflows Sinn, dort allerdings muss der Service-Provider sehr viel enger als bisher in die Prozesse seiner Kunden integriert werden. Selbstverständlich gilt dasselbe im übertragenen Sinn auch für die Erbringung von ITC-Services in Eigenleistung. Es macht absolut keinen Sinn mehr, als Utility im Grid verfügbare Standardleistungen noch selbst zu produzieren, soweit man mit diesen nicht selbst an den Markt gehen will.

Die Diskussion über die Vor- und Nachteile eines klassischen Outsourcing von ITC-Services wird sich in Zukunft also in ganz anderer Form stellen als heute. Für Basisleistungen wird diese Diskussion beim Übergang auf ein Service-Grid auf jeden Fall aufgrund der technischen Gegebenheiten obsolet. Es geht in Zukunft nur noch um die Frage, ob die Integration und Betreuung der Geschäftsprozesse ausgelagert oder selbst erledigt wird.

2.6 Status quo: Ein Application-Grid für IT-Services im DLR

Um die eben beschriebene Vision eines Service-grids Realität werden zu lassen, sind noch mehrere Jahre Entwicklungsarbeit im technischen, betriebswirtschaftlichen und juristischen Bereich notwendig. Diese Arbeiten werden teilweise im internationalen Rahmen (EU IST FP 6) und teilweise national (D-Grid Initiative des BMBF) getrieben und gefördert. Für eine Forschungseinrichtung wie das DLR (Deutsches Zentrum für Luft- und Raumfahrt) macht es jedoch Sinn, bereits heute Infrastruktur und Service-Modell auf eine Basis zu stellen, die evolutionär in Richtung eScience weiterentwickelt werden kann.

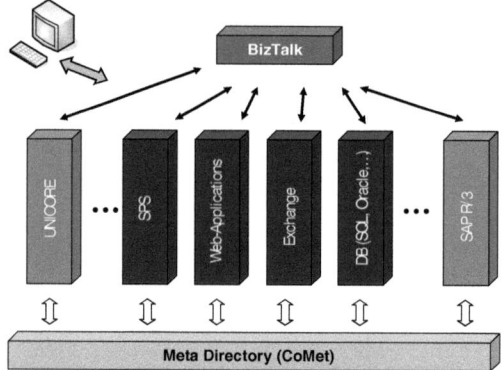

Abb. 4: SOA beim DLR. *[T-Systems]*

Methodik und Status

Neben den notwendigen Entwicklungsarbeiten ist es für Forschungsorganisationen wichtig, die Funktionalitäten von eScience, die bereits heute nutzbar sind, zunächst in Pilotprojekten, später in Produktion einzusetzen und auf der Basis der gewonnenen Erfahrungen die weitere Entwicklung zu beeinflussen. Als Einsatzfelder bieten sich bestimmte Klassen technisch-wissenschaftlicher Anwendungen genauso an wie Informationsdienste und betriebswirtschaftliche Systeme. Der Zugriff auf externe Ressourcen zur Vermeidung eigener Investitionen ist bereits in vielen Fällen möglich, wobei für eine wirtschaftliche Gesamtbetrachtung die Netzkomponente nicht außer Acht gelassen werden darf.

Im Rahmen des Projektes Service-Orientierte Architektur (SOA) werden im DLR zunächst alle Services, die den sog. IT-Backbone bilden, d.h. die gesamten Informations- und Messaging-Dienste, in diese, auf Standard Webservices beruhende Architektur migriert. Es folgen die betriebswirtschaftlichen Informationssysteme samt zugehörigen Geschäftsprozessen. Im Bereich der technisch-wissenschaftlichen Systeme wird schwerpunktmässig der Zugang zu Rechenressourcen für die Simulation in die serviceorientierte Architektur überführt. Dabei kommt die im Rahmen von mehreren BMBF und EU Projekten entwickelte Umgebung UNICORE (Uniform Access to Computing-Resources) zum Einsatz. Dreh- und Angelpunkt des SOA-Konzeptes im DLR ist die Metadirectory Umgebung CoMet (Corporate Metadirectory) (s. Abb. 4).

2.7 Ein Modell der zweiten Generation für ASP

Für ITC-Dienstleister und Nutzer ist die Idee, Anwendungen über ein WWW-Portal allgemein anzubieten und zu nutzen, sehr attraktiv. Der Service-Anbieter kann seine Ressourcen besser auslasten, der Anwender braucht nicht zu investieren.

Für die Softwareanbieter ist dieses Modell dagegen eher problematisch, da es nicht nur die Anzahl verkaufter Lizenzen senkt, sondern wegen des höheren Nutzungsgrades der Einzellizenz auch den Support-Aufwand pro Lizenz erhöht. Bei diesem Modell entsteht also eine Konkurrenzsituation zwischen Software-Anbietern und Service-Providern.

Ein Vorschlag, wie diese Konkurrenzsituation umgangen werden kann, wurde im Rahmen des von der EU geförderten Projektes UniGrids von T-Systems vorgeschlagen. Es orientiert sich an der Entwicklung im Bereich der Internet Service-Provider und geht von einer strikten Trennung von Content und Infrastruktur aus.

Geschäftspartner des Endnutzers wird also in diesem Modell der Anbieter von Software oder von Anwendungsleistungen. Der

Betreiber der Infrastruktur kümmert sich um die Bereitstellung von Ressourcen im Compute- und Netzbereich, um Sicherheit, sowie um Accounting und Billing gegenüber allen beteiligten Parteien.

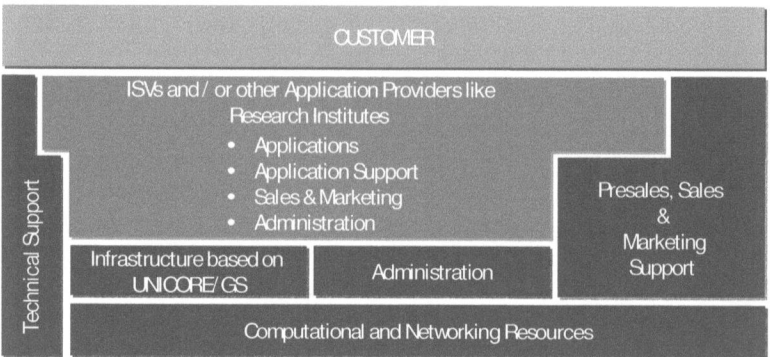

Abb. 5: Zukünftiges ASP-Modell [T-Systems-SfR]

2.8 Zusammenfassung und Ausblick

Um die Grid-Technologie hat sich bereits eine sehr hohe Dynamik entwickelt. Ziel ist der weltweite Zugriff auf Ressourcen und Dienstleistungen. Die Standards für die Architektur sind bereits größtenteils gesetzt, die Realisierung der Vision eines Service-Grid für ingenieurwissenschaftliche Geschäftsprozesse steht jedoch erst am Anfang.

Neben den notwendigen Entwicklungsarbeiten ist es heute extrem wichtig, die Funktionalitäten, die bereits heute nutzbar sind, zunächst in Pilotprojekten, später in Produktion einzusetzen und auf der Basis der gewonnenen Erfahrungen die weitere Entwicklung zu beeinflussen. Als Einsatzfelder bieten sich bestimmte Klassen technisch-wissenschaftlicher Anwendungen genauso an, wie Informationsdienste und betriebswirtschaftliche Anwendungen. Der Zugriff auf externe Ressourcen zur Vermeidung eigener Investitionen ist bereits in vielen Fällen möglich wobei für eine wirtschaftliche Gesamtbetrachtung die Netzkomponente nicht außer Acht gelassen werden darf.

3 Ökonomische Bewertung der Dienstauswahl in Service-Netzen

T. Eymann, M. Reinicke und W. Streitberger

Einfache IT-Dienstleistungen werden im Konzept des On-Demand bzw. Grid Computing an externe Anbieter ausgelagert. Dieser bedarfsabhängige Bezug von Rechenleistung wird als Chance zur Beseitigung von Ineffizienzen und zur Kostenreduktion gesehen. Der vorliegende Beitrag beschäftigt sich insbesondere mit Verfahren zur Dienstauswahl, über die in den vorliegenden Konzepten solcher serviceorientierter Architekturen keine Aussagen bezüglich Anwendbarkeit und Performanz gefällt werden. Er vergleicht die existierenden Ansätze mit einem koordinatorfreien – auf ökonomischen Prinzipien beruhenden – Ansatz anhand ökonomischer Metriken. Mittels einer Simulation werden unterschiedliche Verfahren in verschiedenen Netzwerkszenarien untersucht und die bessere Anpassungsfähigkeit der koordinatorfreien Variante an Dynamik und Knotendichte des Netzwerks gezeigt.

3 Ökonomische Bewertung der Dienstauswahl in Service-Netzen

3.1 On-demand Computing in der serviceorientierten Architektur (SOA)

Die Unternehmen Sun Microsystems, IBM und HP arbeiten seit einigen Jahren in ihren Softwarearchitekturen daran, Kunden IT-Dienstleistungen bei Bedarf („on-demand") anzubieten. Das Geschäftsmodell des On-demand Computings (ODC), erlaubt es Unternehmen, Computerressourcen nur bei Bedarf zu nutzen und zu bezahlen. On-demand Computing ist eine Kombination zweier Vorteile. Erstens ist es ein Dienstzugriffsmodell, bei dem Unternehmen über die IT-Infrastruktur eines externen Dienstleisters Zugang zu zusätzlichen Ressourcen erhalten, um Bedarfsspitzen abzufangen. Zweitens wird mit Hilfe eines Zählwerterfassungsverfahrens ein „Pay as you go"-Bezahlmodell ermöglicht, das die Abrechnung nur derjenigen IT-Dienste und Ressourcen, die das Unternehmen auch tatsächlich nutzt, erlaubt. Somit wird es Unternehmen einfacher gemacht, bedarfsgerecht zu kalkulieren [1, 2, 3].

Grid Computing ist eine Form der Virtualisierung, die dem On-demand Ansatz folgt. Grid Computing beschreibt die Bündelung einer großen Zahl von Rechnersystemen, deren aggregierte Prozessorleistung einen virtuellen Computer ergibt, der besonders rechenintensive Aufgaben durchführen kann. Dabei spielt es keine Rolle, an welchem physikalischen Ort diese Leistung erbracht wird. Solche Systeme werden derzeit genutzt zur Berechnung von z.B. Klimaveränderungen, zur Krebstherapie oder der Entschlüsselung des Genoms [4]. Außerdem ermöglicht das ODC Unternehmen, brachliegende Kapazität zu nutzen. So lassen einige Unternehmen die Desktop-PCs ihrer Mitarbeiter nachts virtuell zusammenschalten, um aufwendige Berechnungen auszuführen. Grid Computing steht für einen Spezialfall des On-demand Computing.

3.1.1 On-demand Computing: Kostensenkung oder Risikoerhöhung?

Aus betriebswirtschaftlicher Sicht sprechen in der Informations- und Kommunikationstechnologie (IuK-Technologie) besonders Kostenargumente für das selektive Fremdbeziehen von eng begrenzten, klar definierten Aufgaben. Für in Eigenregie durchgeführte Tätigkeiten muss das Unternehmen die notwendige Ausrüstung und das erforderliche Know-how besitzen bzw. erwerben und weiterentwickeln. Dies kann nur mit organisatorischen und humankapitalintensiven Anstrengungen erreicht werden, was aber besonders kleinen und mittelständischen Unternehmen (KMU), deren Kernkompetenz meist nicht im IT-Bereich liegt, schwer fällt. Auch mit den rasanten technologischen Neuerungen Schritt zu halten, ist mit Aufwand versehen, der bequem externalisiert werden könnte. In besonderem Licht stehen besonders die periodisch anfallenden Kosten, die bei eigener Bereitstellung unabhängig von der Nutzung der Dienste anfallen. Hierunter fallen unter anderem Raum-, Personal- und Instandhaltungskosten, die einen immensen Fixkostenblock darstellen.

Bei Fremdvergabe von Aufträgen entsteht jedoch eine Unsicherheit über den entfernten Ablauf des Prozesses, bedingt durch einen inhärenten Verlust der Kontrolle, welcher dem Effekt der möglichen Kosteneinsparung gegenüber steht. Abgeleitet aus der Principal-Agent-Problematik kann ein nicht steuerbares bzw. nicht anreizkompatibles Verhalten des Fremdanbieters zu mangelnden Sicherheitsniveaus und Datenschutzproblemen führen und damit erwirtschaftete Marktpositionen ernstlich gefährden. Nicht selten wird bei den Anbietern an diesen, im Vorhinein nicht kontrollierbaren, Leistungseigenschaften – z.B. der dauerhaften Verfügbarkeit – gespart, um die Angebotspreise zu senken und damit attraktivere Verhandlungspositionen einzunehmen. Es schließen sich technische und ökonomische Problemfelder an, die dem Effekt der Kostensenkung massiv entgegen wirken können, so dass schließlich ein „Insourcing" sinnvoller ist. Hier sind geeignete Mechanismen gefragt, die das Risiko einer Auslagerung kalkulierbar und damit den gesamten Prozess effizient machen.

Generell sind On-demand Architekturen also wirtschaftlichen Problemen ausgesetzt. Diese Probleme können jedoch meist auf das Nichtvorhandensein von marktlichen Mechanismen zurückgeführt werden [5]. Zur Analyse der noch einzuführenden, wirtschaftlichen Eigenschaften, wird davon ausgegangen, dass vollständig funktionierende Märkte nur dann auf Dauer Erfolg versprechend sind, wenn sich alle betriebswirtschaftlichen Transaktionsphasen auf den zu untersuchenden Markt abbilden lassen. Als Transaktion als solche wird die Vereinbarung und Regelung über den Tausch aufgefasst. Aufgrund asymmetrischer Informationsverteilung fallen dabei jedoch Probleme (und Kosten) in den einzelnen Phasen an [6], die auf klassischen Märkten durch (einen Markt voraussetzende) Institutionen gelöst werden können. Diese Institutionen sind im ODC jedoch nicht vorhanden und daher bedarf es zu einer Outsourcing-Entscheidung einer genaueren Analyse der Kosten.

3.1.2 Unterstützung der Transaktionsphasen durch Standardisierung

Bei On-demand Verträgen ist die Betrachtung der Transaktionskosten von essentieller Bedeutung. Die externen Transaktionskosten repräsentieren die Kosten der Marktnutzung. Die Höhe dieser Kosten hängt in besonderem Maße davon ab, wie spezifisch die zu erbringende Leistung ist und wie oft diese vom gleichen Anbieter erbracht wird. Denn über einen langen Nutzungszeitraum amortisieren sich dann die Kosten der Marktnutzung. Eine geringe Spezifität der auszulagernden Aktivität erlaubt dabei die preisgünstige Externalisierung. Bei hoher Spezifität dagegen ist es schwierig, die Leistung zu beschreiben und zu bewerten. Sind die Leistungen hoher Spezifität jedoch bereits im Vorfeld weitestgehend standardisiert, kann dem Effekt geeignet entgegengewirkt werden. Die serviceorientierte Architektur (SOA) versucht hier, diese Spezifität durch eine Standardisierung zu reduzieren und ist im Wesentlichen eine Sammlung von Diensten innerhalb eines Netzwerks. Diese Dienste haben die Fähigkeit, miteinander zu kommunizieren und Daten auszutauschen. Die Nachrichteninhalte können entweder einfache Daten zur direkten Nutzung durch den Anwender oder Verwaltungsdaten zur Koordination zweier oder mehrerer Dienste sein. Serviceorientierte Architekturen sind jedoch keine neue Erfindung, die erste SOA wurde etwa mit der Nutzung von DCOM oder Object Request Brokern (ORBs) auf der CORBA Spezifikation realisiert [7].

Seit den letzten Jahren tauchen im Zusammenhang mit der SOA immer wieder die Begriffe *Web Services* und *SOAP* auf. Diese Begriffe dienen der Spezifizierung der Dienste und Protokolle in der SOA und zielen damit auf eine Verringerung der Spezifität und damit auch der asymmetrischen Informationsverteilung in der ersten Transaktionsphase, der Anbahnungsphase, ab: Für die Dienstsuche beschreibt die WSDL (Web Service Description Language) [8] in einem XML-Format den Dienst mit bestimmten Kriterien. Zusammen mit UDDI (Universal Description, Discovery and Integration) , den gelben Seiten der Web Services, kann ein Informationsdefizit aufgebrochen werden und ein Dienst aufgefunden werden [9].

3.1 On-demand Computing in der serviceorientierten Architektur (SOA)

SOAP [10] wiederum spezifiziert die Nachrichten, die zwischen den Applikationen auf Seiten der Nachfrager und Anbieter ausgetauscht werden. So können die Hürden der Kommunikation verschiedener Anwendungen, die womöglich auf verschiedenen Betriebssystemen laufen, überwunden werden. Hier werden demnach die Phasen Verhandlung und Abwicklung unterstützt.

Seit Beginn des Jahres 2004 verfolgen die Firmen IBM, HP und Sun zusammen mit der Globus Foundation das Ziel, Web Services und Grid Computing durch eine Erweiterung der Web Services Spezifikationen um Web Service Notifications und ein Web Service-Resource Framework (WSRF) zusammenzufassen [11, 12]. WS-Notification ist eine Spezifikation zum so genannten Einleiten eines Web Service Ereignisses; WS-Resource Lifetime erlaubt Betreibern von Web Services das Setzen einer Zeitspanne in der die Definition einer Ressource gültig ist; WS-Resource Properties definieren dagegen, wie solche Daten abgefragt und geändert werden können. WS-Notification und das WSRF erlauben eine standardisierte Infrastruktur für Geschäftsanwendungen und Grid Ressourcen. Diese Spezifikationen schaffen ein „publish and subscribe messaging model" und damit die Fähigkeit, Ressourcen zustandsabhängig zu modellieren [12].

Die SOA ist also ein Ansatz, über standardisierte Nachrichtenprotokolle und Nachrichtenspezifikationen, Transaktionskosten zu senken und leistet damit einen Beitrag, Unsicherheit in den ersten Transaktionsphasen zu reduzieren. Dennoch bleiben Fragen unbeantwortet. Etwa wie (inhaltlich gleiche) Dienste aus einer Liste aufgefundener Dienste ausgewählt werden sollen. Darüber werden in der SOA keine Aussagen getroffen. Mit diesem Problem wird sich der Beitrag im Folgenden beschäftigen. Für die weitere Argumentation wird insbesondere die These aufgestellt, dass die Wahl des Dienstes einen Einfluss auf das Funktionieren bzw. die Leistungsfähigkeit der SOA hat. Bevor jedoch auf die Dienstauswahl als solche eingegangen wird, soll zunächst erläutert werden, wie die Leistungsfähigkeit einer SOA beschrieben und gemessen werden kann.

3.1.3 Technische und ökonomische Bewertungsmetriken

Zur qualitativen Bewertung der Leistungsfähigkeit einer serviceorientierten Architektur tragen mehrere technische und ökonomische Merkmale bei. Die ökonomischen Größen sind meist aus den technischen Merkmalen ableitbar.

Zu den technischen Kriterien gehören z.B.

- die Allokationsrate, die die Anzahl der Kontrakte im Verhältnis zu den gestarteten Dienstanfragen misst;
- die Dauer zwischen dem Starten der Dienstanfrage bis zur erfolgreichen Zuweisung, die durch die Response Time (REST) beschrieben wird und
- die Anzahl der notwendigen Kontrollnachrichten.

Diese Metriken sind, wie bereits angedeutet, im Rahmen der Betrachtung in ökonomische Metriken überführt worden; eine hohe Allokationsrate führt etwa zu einer höheren Anzahl von Kontrakten und damit zu einer Verbesserung des Gesamtnutzens, da die Bedürfnisse der Akteure häufiger befriedigt werden können. Die Allokationseffizienz und die Wartezeit, bis auf den Dienst zugegriffen werden kann, können mit den Kosten des Ausfalles des Netzes bzw. den Risikokosten für ein nicht zeitgerechtes Übertragen des Dienstes assoziiert werden. Eine hohe Wartezeit führt dazu, dass das Risiko in Kauf genommen werden muss, den Dienst nicht am Wunschverfügbarkeitstermin zu erhalten. Niedrigere Werte der Response Time (REST) bzw. der Kommunikationskosten deuten also einen Nutzenzuwachs an.

Überdies erlaubt das Heranziehen dieser Kriterien die Fähigkeit, auf exogene Schocks (Ausfall von Links, Knoten und Diensten; Nachfragespitzen) zu reagieren, zu erfassen: Ein System, das sich schneller erholt, ermöglicht höhere Nutzengewinne als ein System, das leicht gestört werden kann und damit aus den Angeln zu heben ist. Können ausreichende Werte der ökonomischen Messgrößen garantiert werden, so lässt das auch den Schluss zu, dass das System in der Leistungsfähigkeit verlässlicher ist und damit Kosten für eine Versicherung des Ausfalls sowohl auf Seiten der Nachfrager als auch auf Seiten der Anbieter reduziert werden können.

Abb. 1 beschreibt eine Kennzahlenpyramide, die die technischen und ökonomischen Metriken aufeinander abbildet. Nutzen wird hier aufgeteilt in Ertrag und Kosten. Bedarfsgerechte Verfügbarkeit, Risiko und Infrastrukturkosten haben entscheidenden Einfluss auf den Ertrag und die Kosten.

Die Verfügbarkeit lässt sich durch Verhandlungs- und Zugangszeiten darstellen, die beide durch die Allokationseffizienz, das Verhältnis von Anfragen und Akzepten bestimmt werden können. Zusätzlich beeinflusst die Entfernung zum Ort der Diensterbringung die Verfügbarkeit. Sie wird durch die Anzahl der Hops quantifiziert.

Versicherungs- und Risikokosten werden über die Standardabweichungen der technischen Metriken gemessen. Eine geringe Standardabweichung steht für ein stabileres System, das vorhersehbare Resultate erbringt.

Die Infrastrukturkosten können durch die Anzahl der Kontrollnachrichten und die Nachrichtengröße charakterisiert werden, die die notwendige Bandbreite zur Herstellung der Allokationen bezeichnen.

Die Abbildung der Kennzahlen wurde bisher nur qualitativ durchgeführt, eine quantitative Analyse, die technische auf wirtschaftliche Metriken abbildet und umgekehrt, ist bisher nicht geschehen. Für den nun folgenden Vergleich der Dienstauswahlverfahren werden daher lediglich die technischen Metriken in Betracht gezogen, die qualitativ auf die ökonomischen Kennzahlen schließen lassen.

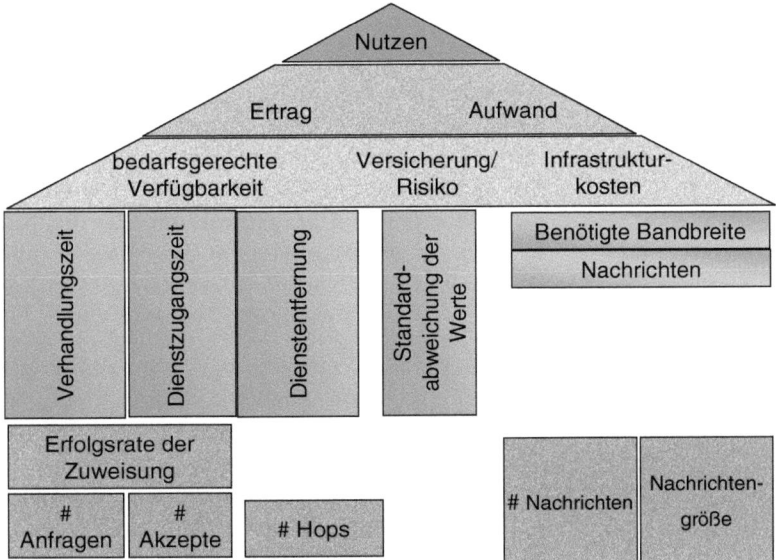

Abb. 1 *Kennzahlenpyramide*

3.2 Dienstauswahlverfahren zur effektiven Selektion

Große und komplexe Computernetze, die verteilte Dienste verschiedener Art bei Bedarf zur Verfügung stellen, werden seit Ende des letzten Jahrhunderts in der serviceorientierten Architektur zusammengefasst. Im Rahmen dieses neuen Ansatzes soll eine große Anzahl von Rechnern lose gekoppelt werden, um eine verteilte Informationssuche, eine parallele Verarbeitung von Aufgaben oder eine persistente Datenspeicherung zu erlauben. Content Distribution Netzwerke [13, 14, 15] oder Peer-to-Peer Netze [16, 17] sind solche Softwarearchitekturen, die Dienste bereitstellen und über Computernetzwerke verfügbar machen. Die Ressourcen sind über einfache Kommunikationsinfrastrukturen – wie dem Internet – verbunden. Diese Netze werden z.B. zum Bereitstellen von Multicast-Diensten für Gruppen [14], zum Speichern extrem großer Datensätze [16] oder zum Ausführen von Anwendungen, die Rechenleistung in der Größenordnung von Gigaflops benötigen [17], eingesetzt. Populäre Anwendungen, die nun auch verstärkt im Interesse der Industrie stehen und das Potential für grundlegende Innovationen bergen, sind etwa das Grid Computing für eine verteilte, parallele Verarbeitung, File Sharing bzw. Persistent Storage Systeme oder Instant Messaging.

Eine wichtige Fragestellung für den Erfolg dieser Anwendungen ist, wie nutzbare Dienste und Informationen aus einer Fülle von Daten im Netz gefunden und extrahiert werden können [18, 19, 20]. Effektive Dienstfindungs- und Dienstauswahlmethoden sind daher notwendigerweise erforderlich. Bei gegebener Komplexität und Dynamik einer Grid Infrastruktur ist die Skalierfähigkeit und die Verwaltung einer hohen Anzahl von heterogenen Ressourcen die überragende Herausforderung für die Nachhaltigkeit. Insbesondere für das Management dieser Netze müssen automatisierte, vorzugsweise selbstorganisierende Lösungskonzepte zu den folgenden Auswahlproblemen bereitstehen:

- Das initiale Verteilen bzw. Neuverteilen von Diensten und Ressourcen im Netzwerk: Aufgrund der Annahme, dass eine spezifische Nachfrage für einen Dienst bzw. Web Service existiert, stellt sich für den Anbieter die Frage, von wo der Dienst physikalisch bereitgestellt werden sollte. Der Ort des Angebots beeinflusst den Ertrag des Dienstes für den Eigentümer maßgeblich und zudem die subjektive Empfindung des Nachfragers (z.B. durch Wartezeiten). Auch die Performanz des Netzes hängt von der Platzierung der Dienste ab. Eine

3.2 Dienstauswahlverfahren zur effektiven Selektion

Veränderung in der Struktur der Nachfragen könnte beispielsweise zu einem notwendigen Neuverteilen der Ressourcen führen. Existierende Ansätze sind etwa Xweb Multicast dampening [21], der OptorSim Replica Optimizer [22] oder DYNAMO [23].

- Das Auffinden von verfügbaren Diensten: Bei Existenz mehrerer redundanter Dienste, die jeweils alleine die Nachfrage befriedigen könnten, stellt sich die Frage, welche dieser Dienste gerade verfügbar sind. Genau diese Instanzen in einer angemessenen Zeitspanne zu finden, ist eine anspruchsvolle Aufgabe (Dienstfindung). Unstrukturierte Verfahren zur Dienstfindung werden bereits in Globus, Gnutella oder Datagrid eingesetzt, garantieren aber das Auffinden nicht. Die aus diesem Grunde entwickelten strukturierten Verfahren stellen ein viel versprechendes Novum dar, da eine Dienstfindung in einer vorhersagbaren Zeit garantiert werden kann. Mit Hilfe von verteilten Datenbanken und Distributed Hash Tables (DHT) sorgen die Netzknoten gemeinsam für ein effizientes Auffinden [24, 25].

- Die Auswahl eines passenden Handelspartners: Aus einer Liste von aufgefundenen Dienstinstanzen muss der Nachfrager bzw. der Resource Broker einen passenden Transaktionspartner wählen, der beide Teilnehmer gleichermaßen befriedigt. Dieser Prozess liegt im Fokus des vorliegenden Beitrages. Die gängigen Grid Systeme bedienen sich meist zentraler Resource Broker zur Auswahl des Dienstes. Verfügbare Ansätze – wie etwa Nimrod/G oder OptorGrid [22] – sind meist mit der Dienstfindung verknüpft.

Im Folgenden werden die verschiedenen Verfahren der Dienstauswahl anhand der Zentralität des Auswahlmechanismus und der Reihungsmethoden zur Allokation kategorisiert.

3.2.1 Zentrale Dienstfindung und -auswahl

Eine typische Form der Realisierung der Dienstauswahl ist die der Verwendung eines zentralen Resource Brokers. Dieser sammelt die ihm zugetragenen Informationen über Nachfrage und Angebot vorhandener Dienste. Die Liste der möglichen Allokationspartner wird zentral angelegt und nach definierten Bewertungskriterien gereiht. Die Akteure (Nachfrager und Anbieter) aktualisieren diese Information in bestimmten zeitlichen Abständen mittels Nachrichten, die ihren derzeitigen Anfragestatus und ihre Verfügbarkeit beinhalten, um den Broker auf einem möglichst aktuellen Informationsniveau zu halten. Denn nur mit aktuellen Daten kann der Broker bzw. Koordinator die Zuordnung fehlerfrei herstellen. Den einzelnen Instanzen wird nach der durch den Broker durchgeführten Zuweisung – die nach einer definierten Anzahl von erhaltenen Daten erfolgt bzw. nach zeitlichen festgelegten Intervallen – nur die Allokationsinformation mitgeteilt. Der Nachfrager tritt dann in den direkten Kontakt mit dem Anbieter, um die temporäre Kooperation zu initiieren. Abb. 2 stellt diesen Verlauf grafisch dar.

Charakteristisches Beispiel ist Condor-G [22, 26]. Condor-G nutzt einen Resource Broker. Jede Nachricht über eine Nachfrage oder ein Angebot wird mit einem so genannten ClassAd versehen. Verfügbare Ressourcen werden gemäß Nutzerpräferenzen, Zuweisungskosten oder Anfangszeitpunkten geordnet. Die ClassAds bestehen unter anderem aus einem speziellen „Requirement and Rank"-Etikett. Das Rank-Kriterium wird genutzt zur Reihung der Angebote. Die verfügbaren Ressourcen und Nutzeranfragen werden zudem in einer bestimmten statischen Frequenz nach ihrem aktuellen Status abgefragt. Nach Ablaufen einer bestimmten Zeitspanne entscheidet der Broker, wer welchen Dienst in Anspruch nehmen darf. Zunächst werden hierzu verfügbare Ressourcen nach Nutzerpräferenzen, Allokationskosten und erwarteten Start- und Bearbeitungszeiten eingestuft. Nach einem Anpassen der Anforderungen werden die Rangfolgen für jedes mögliche Paar gebildet, aufgrund derer über eine Heuristik die Zuweisung erfolgt [27, 28]. Der Broker sendet die Allokationsdaten an die Peers. Diese wickeln die Transaktion ohne Zutun des Brokers ab. Weitere, ähnliche Verfahren finden sich unter [29, 17, 14].

3.2 Dienstauswahlverfahren zur effektiven Selektion

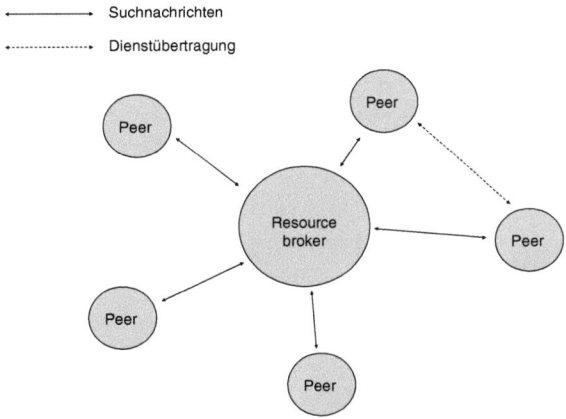

Abb. 2: *Ablauf zentraler Dienstfindung und -selektion*

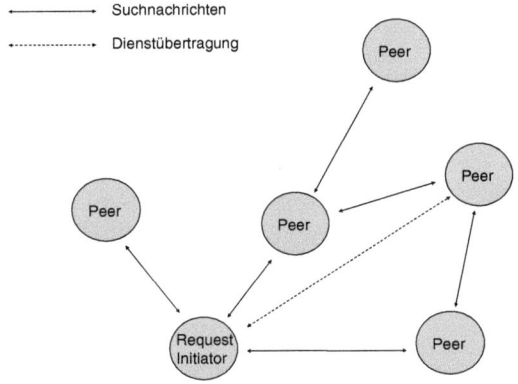

Abb. 3: *Ablauf dezentraler Dienstfindung und -selektion*

Wird der zentrale Broker in Netzen eingesetzt, die nichtstatischer Natur sind, deren Akteure also nicht konstant erreichbar und verfügbar sind, kommt es zwangsläufig zu Allokationen zwischen Handelspartnern, die nicht mehr verfügbar sind, bzw. zu theoretisch besseren Allokationsmöglichkeiten durch das Hinzukommen von neuen Akteuren, sei es durch Ausfälle bzw. Aktivieren von neuen Knoten und Links oder bewusste Nutzereingriffe. Diese suboptimalen Zuweisungen nennt man Fehlallokationen, da sie unter Ausschluss globaler Information den Optimalitätsanspruch nicht gewährleisten können. Um Fehlallokationen effektiv zu vermeiden, ist es notwendig, dass die Ressourceninhaber ihre Verfügbarkeit ständig überprüfen und die beim Resource Broker liegenden Informationen aktualisieren. Die Endnutzer speichern gesendete Anfragen in Warteschlangen und aktualisieren diese Liste kontinuierlich.

Für optimale Ergebnisse ist demnach eine globale Sicht auf das System erforderlich, was jedoch problematisch ist [28]. Im Folgenden sind die Erfordernisse für ein idealisiertes Zuweisen aufgeführt:

- Erstes Erfordernis der zentralen Kontrolle ist, dass der Zustand des Netzes sich zwischen dem Beginn und dem Ende des Berechnungs- bzw. Allokationsprozesses nicht ändern darf. Netzwerke sind jedoch sehr dynamisch und ändern schnell ihren Zustand; Angebot und Nachfrage verschieben sich sehr häufig und Ausfälle [30] treten bei gleichzeitigem Hinzukommen neuer Dienste [31] regelmäßig auf. Dynamische Netze benötigen hingegen eine fortlaufende Aktualisierung des Koordinationsmechanismus, die die ständigen Umweltveränderungen reflektiert.

- Eine zweite Eigenschaft ist, dass der Koordinator über globales Wissen in Bezug auf den Zustand des Netzwerkes verfügen muss. Dies wird meistens durch bestimmte Zeitintervalle zwischen dem ersten Empfang der Statusnachrichten der entfernten Ressourcen bzw. Nachfragen und dem Empfang der letzten Nachricht bestimmt. Bei einem großen Durchmesser des Netzes ergeben sich hohe Latenzzeiten für die Nachrichten, die bei Eintreffen evtl. nicht mehr die aktuellen Daten beinhalten.

- Drittens stellt der Koordinator als solches ein Problem dar. Da Angebots- und Nachfragenachrichten durch das Netz zum zentralen Koordinator geleitet werden müssen, wird dem System unnötiger Datenverkehr aufgebürdet. Solange die Kontrolldaten von der Größe her nur klein im Vergleich zu den Anwendungsdaten sind, ist das weniger problematisch, doch dann, wenn das zentrale Prinzip auf weitere Anwendungsge-

3.2 Dienstauswahlverfahren zur effektiven Selektion

biete angewandt wird, werden erhebliche Kapazitäten in Anspruch genommen, die zu Lasten des Systems gehen [32].

3.2.2 Auswahl der Dienste beim Konsumenten (dezentrale Dienstauswahl)

Eine Alternative zum vorgestellten, zentralen Resource Broker versprechen dezentrale Mechanismen, welche die Entscheidung auf den anfragenden Client verschieben. Typische Realisierungsformen finden sich in Peer-to-Peer Netzen, für die Gnutella [33] ein charakteristisches Beispiel ist. Gnutella nutzt das Fluten des Netzwerkes zur Dienstfindung: Nachdem ein Client eine Anfrage initiiert hat, werden Nachrichten an die Nachbarn verteilt. Auch diese fragen ihre Nachbarn an. Ein Time-to-live Parameter (TTL) begrenzt dieses Ausbreiten von Suchnachrichten und damit den Suchraum unter der optimistischen Annahme, dass der Dienst gefunden wird [19]. Über alle aufgefundenen, auf die Anfrage passenden Dienste wird der Client mittels einer Nachricht, dem so genannten QueryHit, informiert. Diese Nachrichten werden außerdem in den Zwischenknoten gespeichert, um zukünftige Anfragen zu beschleunigen. Abb. 3 zeigt den Verlauf.

Suchverfahren, wie etwa Chord, Pastry oder CAN (Content Addressable Network), nutzen eine Distributed Hash Table (DHT), um dieses Fluten zu umgehen. Die Suche wird in diesen Verfahren über alle Knoten verteilt und ein Auffinden kann garantiert werden [24].

Diese unterschiedlichen Suchprozesse führen jeweils dazu, dass eine Liste der verfügbaren Dienste beim Client generiert wird. Der Client muss nun mit seinem eigenen, lokalen Wissen einen Dienst aus der Liste auswählen. Diese Auswahl trifft er unter Unwissenheit von Netzwerktopologie, etwaigen Flaschenhals- und Knappheitssituationen. Aus diesem Grunde ist die Selektion des Dienstes von besonderer Wichtigkeit, da sie über den Erfolg der Transaktion entscheidet. Zum Zwecke der Entscheidung wird die Wahl mit Hilfe einer gereihten Liste durchgeführt.

3.2.3 Ordnungsmethoden

Die Reihung der gefundenen Dienste hat eine signifikante Auswirkung auf die Güte des Matchings. Sowohl der Resource Broker in der zentralen Variante, als auch der Nachfrager im dezentralen Szenario generieren bei der Suche eine Liste. Jeder Treffer muss anhand eines Verfahrens bewertet und in eine Reihe eingeordnet werden, die zur Weiterverwendung genutzt wird. Wie auch bei der Nutzung von Suchmaschinen im Internet haben diese Listen einen entscheidenden Einfluss darauf, welcher Link bzw. Dienst gewählt wird. Als Analogon dienen Suchmaschinen im World Wide Web, deren Leistung darin besteht, dem Nutzer auf den ersten Plätzen Inhalte zu präsentieren, die seiner zuvor eingegebenen Suche entsprechen und das Informationsdefizit beheben. Für die Reihung existieren unterschiedliche Methoden, die im Folgenden kurz erläutert werden.

3.2.3.1 Ordinale Reihung

Die Auswahl eines Dienstanbieters aus einer Liste von Kandidaten wird in den meisten Anwendungen durch eine Reihung nach technischen Merkmalen durchgeführt, z.B. anhand des Kriteriums „Antwortzeit". Die möglichen Dienste werden nach Antwortzeit aufsteigend sortiert und der Nutzer wählt den Dienst an erster Stelle aus. Ist dieser Dienst jedoch nicht mehr verfügbar, so bedient er sich des nächsten Dienstes in der Liste, solange bis ein Kontrakt zustande kommt. Ist kein Kontrakt zu Stande gekommen, die Liste aber abgearbeitet, so muss eine neue Anfrage gestartet werden.

Diese Reihung nach technischen Kriterien hat zum Nachteil, dass die Unterschiede zwischen den Plätzen der Liste nicht durch Beträge dargestellt werden, die den absoluten Nutzenverlust von first-best auf second-best darstellen. Es kann daher nicht gesagt werden, wie deutlich eine Verschlechterung bei Ausfall eines bevorzugten Dienstes ist. Eine Lösung verspricht daher die Verwendung ökonomischer Prinzipien, die im Folgenden dargestellt werden.

3.2.3.2 Ökonomische Reihung durch interne Preise

Für die Nutzung ökonomischer Prinzipien müssen das Netz und die Güter jedoch wertorientiert betrachtet werden, was dazu führt, dass ein Preis für alle beziehbaren Netzressourcen existieren muss. Das erlaubt die Berechnung eines Nutzens auf Seiten des Nachfragers, der entsprechend den Angebotspreis mit einem potentiellen Nutzen bei Vertragserfüllung vergleichen kann und daraus seinen Nutzenzuwachs berechnen kann.

3.2 Dienstauswahlverfahren zur effektiven Selektion

In einigen Projekten, wie etwa GridBus oder Nimrod/G, wird die Ergebnisliste über die Bewertung von Nutzenzuwächsen sortiert. Damit können die Unterschiede in der Reihenfolge eindeutig in Nutzendifferenzen dargestellt werden. Der Nimrod/G Resource Broker ist ein Planungs- und Steuerprogramm, das für Dienstauffindung, -auswahl und -verteilung zuständig ist. Während erste Versionen noch ausschließlich fristenbasiert allokiert haben, entscheiden aktuelle Versionen mit Hilfe ökonomischer Prinzipien: Die implementierte GRACE Infrastruktur innerhalb des Brokers erlaubt es, um Ressourcen zu handeln – ähnlich wie in einem Markt. Die Entscheidung basiert damit auf Kosten und Preisen, aber weiterhin auch auf Zugangsnormen und Fristen [34].

Diese ökonomischen Prinzipien werden z.B. von katallaktischen Informationssystemen unterstützt. Das Implementieren dieser Informationssysteme nutzt Methoden sowohl aus der Agententechnologie, als auch der Ökonomie, so genannte agentenbasierte „computational economics" [35]. Autonome Softwareagenten verhandeln untereinander, um ihren Nutzen zu maximieren und passen ihre Verhandlungsstrategie unter Verwendung von maschinellen Lernverfahren stetig an (Evolutionäre Algorithmen, numerische Optimierung, z.B.: Nelder/Mead's Simplexmethode [36], hybride Methoden, z.B. das VID Modell von Brenner [37]). Die Signalwirkung von Preisen führt zu einer konstanten Anpassung des Gesamtsystems und verbreitet Veränderungen in der Knappheit von Ressourcen innerhalb des Systems. Die sich ergebenden Muster sind vergleichbar mit denen, die in Marktexperimenten mit menschlichen Akteuren beobachtet werden [38, 39, 40]. Eine erfolgreiche Anwendung der Katallaxie in verteilten Allokationsverfahren verspricht den Vorteil einer flexiblen Struktur und eines inhärenten, parallelen Verarbeitens im Vergleich zu einem zentralen, auktionatorbasierten Ansatz. Ein Beispiel zur Simulation einer Wertschöpfungskette ist das Multiagentensystem AVALANCHE [41].

Von der Verwendung ökonomisch rationaler Prinzipien in Dienstauswahlverfahren wurde bisher jedoch meist abgesehen. Open Agoric Systems waren ein früher Versuch, marktliche Prinzipien in Computer Systeme zu implementieren. Aufgrund mangelnder Notwendigkeit für reale Anwendungen und fehlender Skalierbarkeit schenkte ihnen die Wissenschaft kaum Beachtung [42]. Die Tauschbörse Mojo Nation [43, 44], die die Ressourcen, wie etwa Speicherplatz, nur über einen Markt anbot, hatte Akzeptanzprobleme einerseits wegen des Fehlens von erprobten Bezahlverfahren für Micropayments und andererseits wegen des Feststellens von Sabotageakten der Nutzer, die sich durch Betrachten Ihres virtuellen Kontos angespornt fühlten, das System auszustechen. Zusammenfassend lässt sich sagen, dass Dienstauswahlverfahren zentralisiert oder dezentral durchgeführt werden können, während sich der Auswahlmechanismus technischer oder rationaler, ökonomischer Prinzipien zum Reihen der Ergebnisse bedienen kann. Bildet man diese Ergebnisse auf eine 2x2 Matrix, so ergibt sich Abb. 4.

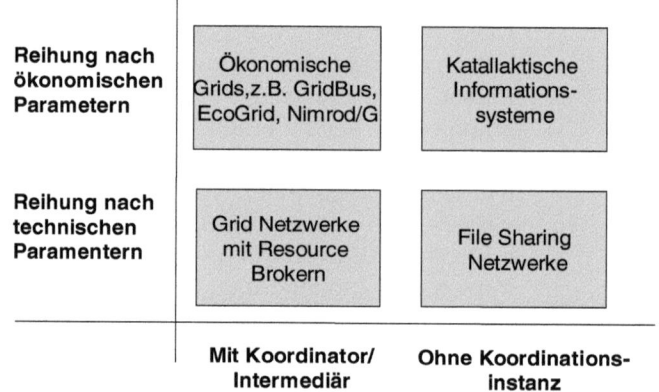

Abb. 4: *Dienstauswahlverfahren: Stand der Wissenschaft*

3.3 Simulation und Evaluation

Die vorgestellten Dienstauswahlverfahren sollen anhand einer Simulation einander gegenübergestellt und für die Einsatzeignung in unterschiedlichen Szenarien technisch und ökonomisch anhand der in Kapitel 3.1.3 vorgestellten Metriken bewertet werden. Zur Vereinfachung wurde auf den Vergleich zwischen den verschiedenen Ordnungsmethoden verzichtet. Es wurden demnach nur die oberen beiden Verfahren aus Abb. 4 zur Simulation herangezogen, die Dienstauswahl mit und ohne Koordinator bzw. Resource Broker. Eine einfache Simulationsumgebung für dieses Vorhaben wurde erstmals im Projekt CatNet [45] entwickelt und seitdem weiterentwickelt und erweitert. Dieses Kapitel soll die formulierte These bzgl. des Einflusses der Dienstauswahl auf die Leistungsfähigkeit der SOA untermauern und einen Vergleich der Verfahren ziehen.

3.3.1 Netzwerkattribute und Hypothesen

Zum Vergleich wurden die zu untersuchenden Verfahren in ein Netzwerk, bestehend aus 106 Netzknoten eingebettet. Auf den 106 Knoten waren 75 Clients (Nachfrager) an den Aussenkanten verteilt, den so genannten Edges, auf allen Knoten konnten dagegen Dienste liegen (Abb. 5).

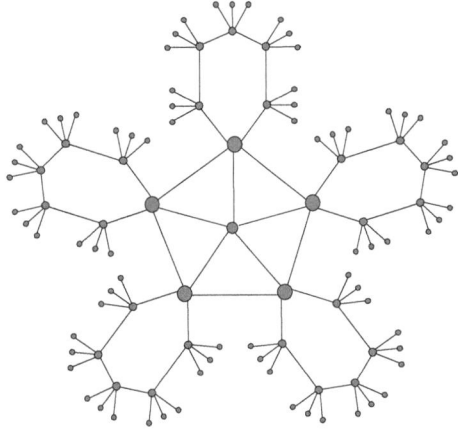

Abb. 5: Netztopologie

Das Netz sollte in verschiedenen Simulationsläufen unterschiedliche Eigenschaften bezüglich der physikalischen Konzentration und der stetigen Verfügbarkeit der Dienste haben, um damit relativ einfach auf existierende Netze abstrahiert werden zu können. Dazu wurden jeweils fünf Stufen der Konzentration und der Verfügbarkeit (bzw. Ausfallwahrscheinlichkeit) der Dienste in den Simulator implementiert. Die Dienste wurden bei Initialisierung nach dem Zufallsprinzip in fünf verschiedenen Konzentrationen verteilt. Folglich wurden für 2 Verfahren jeweils 25, also insgesamt 50 Simulationsläufe durchgeführt (Abb. 6).

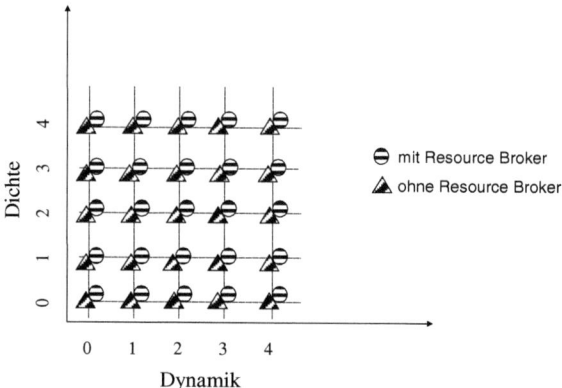

Abb. 6: *Testszenarien mit den Dimensionen Dichte und Dynamik*

Die Dichte bzw. Konzentration der Dienste kann in Netzen sehr unterschiedlich sein. In so genannten Content Distribution Netzen (CDN) zur Verteilung von Web Content liegen die Dienste physikalisch sehr konzentriert, da die Webseiten von nur wenigen Servern millionenfach an die Clients verteilt werden. Man spricht davon, dass die Dienste hier sehr dicht konzentriert sind. Für die Simulation wird dieses Szenario mit Dichte 0 repräsentiert. Dagegen sind Dienste in Ubiquitous Computing Szenarien meist weit verteilt und an einem physikalischen Ort liegt nur ein Dienst physikalisch. Dichte 1 bis 4 stellen die Übergänge zur Veranschaulichung des Verhaltens der in Kapitel 3.1.3 vorgestellten Metriken dar.

Sind ein Netz und seine Knoten über den Beobachtungszeitraum statisch in der Verfügbarkeit der Dienste und Links, so spricht man von statischen Netzen. Diese Netze liegen vor, wenn die Teilnehmer Ihre Verfügbarkeit durch organisatorische Maßnahmen garantieren können, z.B. Unternehmensnetzwerke. Diese werden in der Simulation mit der Dynamikstufe 0 repräsentiert. Auf der anderen Seite gibt es Netze, deren Dienstressourcen nur kurze Zeit verfügbar sind und neue Dienste hinzukommen. In Peer-to-Peer File Sharing Anwendungen kann dieses Verhalten etwa beobachtet werden. Die Stufen dazwischen sollen einen Übergang dieser beiden Extrema herstellen.

Innerhalb eines Experimentes senden die 75 Clients insgesamt 2000 Nachfragen, die Dauer der exklusiven Dienstbeanspru-

3.3 Simulation und Evaluation

chung, also des Vertragsverhältnisses, betrug konstant 50ms, danach war der Dienst wieder frei verfügbar. Die Abstände der Anfragen eines einzelnen Clients wurden so gesteuert, dass das System ständig unter Last steht (Ergebnisse der Experimente zur Bestimmung der Eingangswerte finden sich unter [45]). Die verschiedenen Eingangswerte für die Dichte und Dynamik sind in Tab. 1 und Tab. 2 dargestellt.

Tabelle 1: Simulationseingangswert für den Parameter Dichte

Dichte	Verteilung der 300 Dienste auf vorhandene Knoten
0	6 Knoten mit je 50 Diensten
1	15 Knoten mit je 20 Diensten
2	25 Knoten mit je 12 Diensten
3	50 Knoten mit je 6 Diensten
4	75 Knoten mit je 4 Diensten

Tabelle 2: Simulationseingangswerte für die Dynamik

Dynamik	Wahrscheinlichkeit des Dienst-ausfalls in % (Messung alle 200ms)
0	0
1	15
2	30
3	45
4	60

3.3.2 Der Netzwerksimulator J-Sim/TCL

Der Unterbau der Simulationsumgebung wird von J-Sim bereitgestellt. J-Sim ist eine objektorientierte, komponentenbasierte Umgebung zur diskreten Simulation von frei definierbaren Netzwerktopologien [46]. Die Netzwerkeigenschaften (Protokolle, Bandbreiten, Verlustraten etc.) können dabei frei bestimmt werden und tragen damit zu einer möglichst wirklichkeitsgetreuen

Simulationsmodellierung bei. Die Simulation findet auf TCP/IP- bzw. UDP/IP-Paketebene statt und lässt vielfältige Freiheitsgrade in der Konfiguration zu. J-Sim kann derartig konfiguriert werden, dass ein spezifisches Anwendungsschichtnetz simuliert wird. Zudem können verschiedene Softwareagenten auf die Knoten aufgesetzt werden, die autonom handeln, und Ihren Nutzen zu maximieren versuchen. In der im Rahmen dieses Beitrages durchgeführten Simulation wurden spezielle Agenten für Kunden, Speicherplatzressourcen und Dienste programmiert und auf die entsprechenden Knoten gesetzt. Die Konfiguration des Netzwerkes erfolgt über Tcl/Tk-Skripte, die dynamisch generiert werden.

3.3.3 Simulationsergebnisse

In diesem Abschnitt finden sich die aktuellen Ergebnisse der Simulation. Die Simulation wurde 200 Mal mit den jeweils gleichen Eingangswerten durchgeführt, was zusätzlich eine anschauliche Übersicht über die Verteilung und Abweichungen der Ergebnisse liefert.

Aus Platzgründen wird ausschließlich auf die Metrik der Allokationsrate (RAE) und die Anzahl der Kontrollnachrichten eingegangen.

3.3.3.1 Allokationseffizienz mit Resource Broker

Abb. 7 demonstriert den Verlauf der Allokationsrate bei Einsatz eines Resource Brokers. Es werden dazu die Mittelwerte der Ergebnisse durch einen nicht ausgefüllten Kreis und das 95%-Konfidenzintervall mit den oberen und unteren Grenzen dargestellt. Während in Dichte 0 und Dynamikstufe 0 knapp 94% der Anfragen befriedigt werden, erkennt man, dass bei Anstieg der Verteilung von Dichte 1 auf 2 ein deutlicher Rückgang erkennbar ist (bis auf 72% in Dichte 2). Dieses Ergebnis liegt auf der Hand, da in statischen Netzen eine vollständige Sicht über das Netz erreicht werden kann und damit die Zuweisung recht nahe am Optimum liegt. Erhöht sich nun die Verteilung der Dienste, so ist dies mit einer Komplexitätssteigerung zu vergleichen: Der Resource Broker hat Probleme mit Objekten die an vielen, verschieden weit entfernten Orten liegen. Erhöht man zusätzlich die Ausfallwahrscheinlichkeit, so ist das Defizit weit deutlicher erkennbar. In Dynamikstufe 4 werden bei Konzentration (Dichte 0)

3.3 Simulation und Evaluation

der Dienste nur 52% der Anfragen befriedigt, in Dichte 2 und Dynamikstufe 4 nur noch 40% der Anfragen.

Die Konfidenzintervalle sind in der Dichte 0 etwas größer wie die in Dichte 4. Offensichtlich sind also die Ergebnisse bei Verteilung der Dienste relativ stabil.

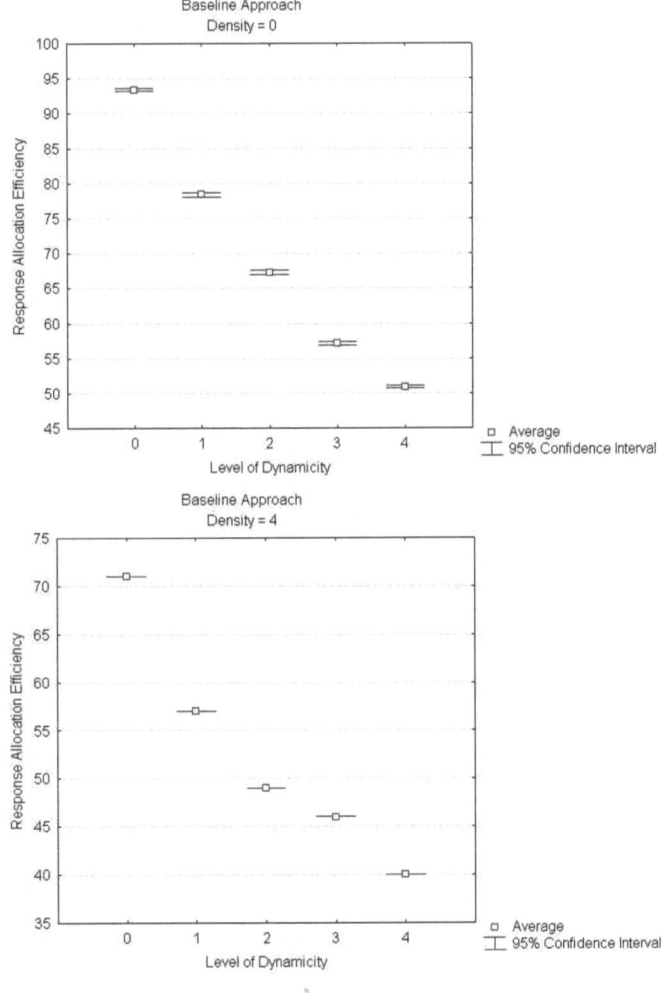

Abb. 7: *Mittlere Allokationsrate und 95%-Konfidenzintervall mit Resource Broker in Dichtestufe 0 (oben) und 4 (unten)*

3 Ökonomische Bewertung der Dienstauswahl in Service-Netzen

3.3.3.2 **Allokationseffizienz ohne Resource Broker**

Abb. 8 zeigt den grafischen Verlauf der Allokationseffizienz bei dezentraler, marktlicher Koordination über die Dynamikszenarien hinweg. Zunächst ist erkennbar, dass die besten Werte erzielt werden für Netze, die eine geringe Ausfallwahrscheinlichkeit (Level of Dynamicity 0) aufweisen. Hier werden Werte von 98% erreicht. Bei ansteigender Dynamik der Dienste fällt die Allokationsrate jedoch schnell ab. Bei einer Ausfallwahrscheinlichkeit von 60% (Dynamik 4) können bei Dichtestufe 0 nur noch 79% der Anfragen befriedigt werden. Auffällig ist, dass bei einer breiteren Verteilung der Dienste, also dem Ansteigen der Dichtestufe, diese Rate nicht so stark fallend ist. In Dichtestufe 4 werden noch 89% der Anfragen erfolgreich zugewiesen und die Transaktion abgewickelt. Somit beträgt das Defizit bei Erhöhung der Ausfallwahrscheinlichkeit unter 6%. Eine breitere Verteilung der Dienste fängt also die steigende Ausfallwahrscheinlichkeit ab; offensichtlich kann der Ansatz unter Abwesenheit eines Resource Brokers besser mit physikalisch verteilten Dienststandorten umgehen. Die Konfidenzintervalle verkleinern sich mit der Verteilung, was für ein stabileres System spricht.

3.3.3.4 **Vergleich der gesendeten Nachrichten auf IP-Ebene**

Abb. 9 und Abb. 10 zeigen die Anzahl der versandten Nachrichten bzw. IP-Pakete zwischen den einzelnen Knoten bzw. Datenübertragungseinrichtungen. Zunächst bemerkt man die recht hohe Zahl von Nachrichten, die einerseits durch das Fluten des Netzes zur Dienstfindung und den Nachrichten zur Verhandlung eines Dienstes bedingt ist, andererseits durch die große Anzahl an Nachrichten zur Aushandlung der Verträge.

Es fällt auf, dass die Variation der Dynamik einen erheblichen Einfluss auf die Menge der Nachrichten hat; auch die Verteilung der Dienste (höhere Dichtestufe; hier von Stufe 1 auf 2) führt zu einem höheren Nachrichtenaufkommen in beiden Ansätzen: Sowohl mit als auch ohne Koordinator steigt die Nachrichtenanzahl an. Während in Dichtestufe 1 die Zahl der Nachrichten mit Resource Broker geringer ist als ohne, wird mit steigender Dichtestufe diese Verhältnismäßigkeit umgedreht. Dies ist begründet durch eine höhere Anzahl an Nachrichten, die dem Koordinator zugeführt werden muss bei Änderung der Verfügbarkeitsstati der Ressourcen. Die dezentralen Verhandlungen neigen dazu, bei

3.3 Simulation und Evaluation

höherer Dynamik mehr Nachrichten zu benötigen, das Volumen der Nachrichten steigt also überproportional, während das dezentrale Verfahren mit unterproportional vielen Nachrichten auskommt.

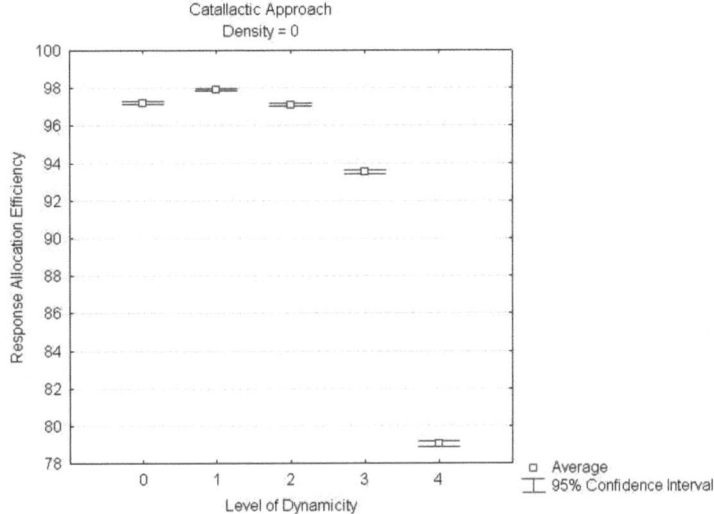

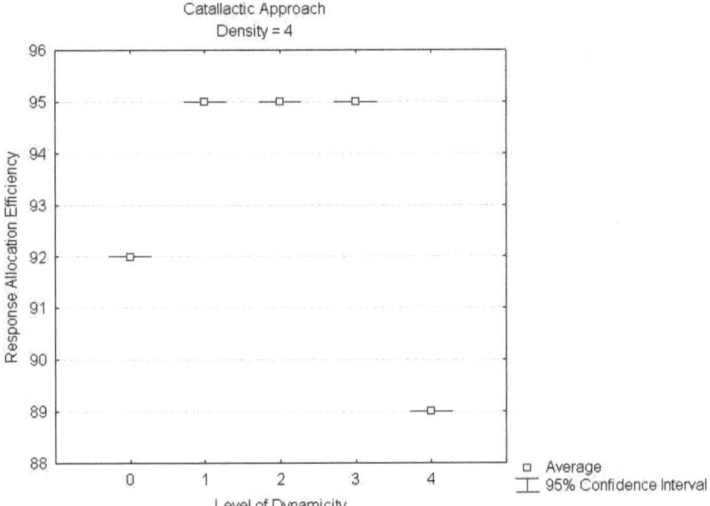

Abb. 8: *Mittlere Allokationsrate und 95%-Konfidenzintervall ohne Resource Broker in Dichtestufe 0 (oben) und 4 (unten)*

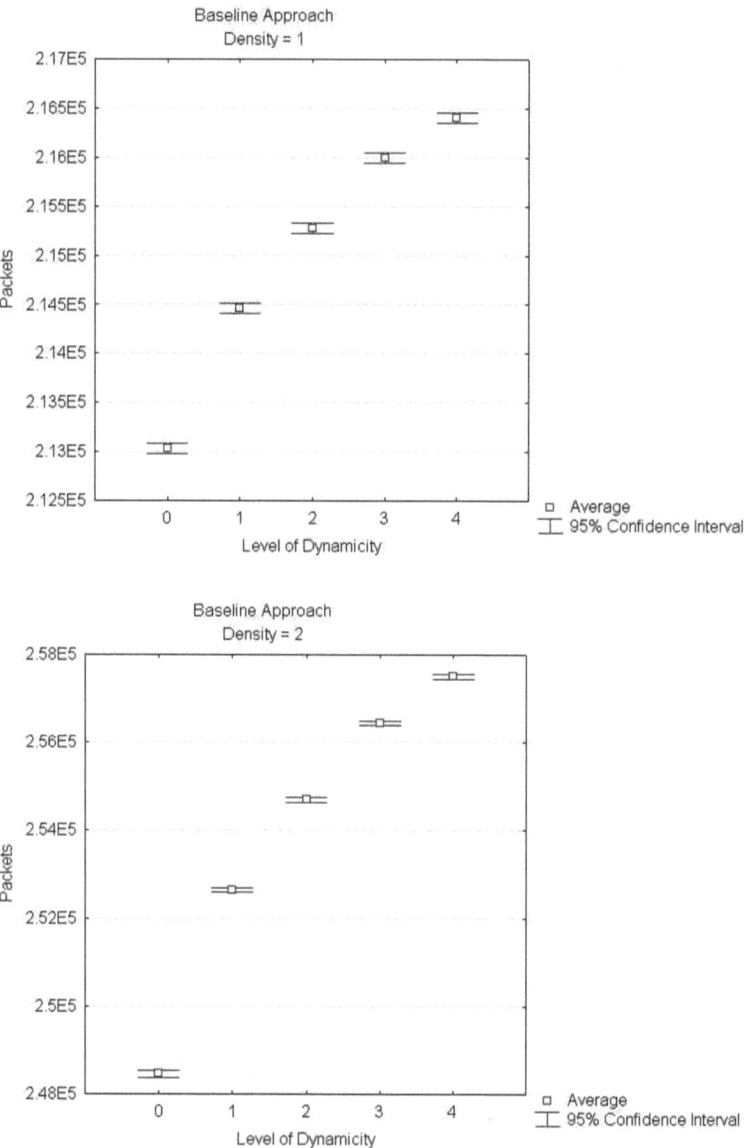

Abb. 9: *Mittlere Anzahl der Pakete und 95%-Konfidenzintervall mit Resource Broker*

3.3 Simulation und Evaluation

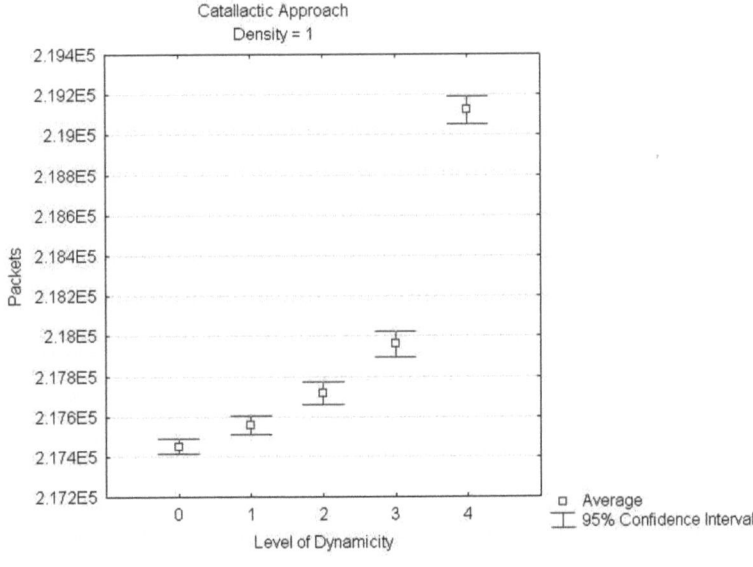

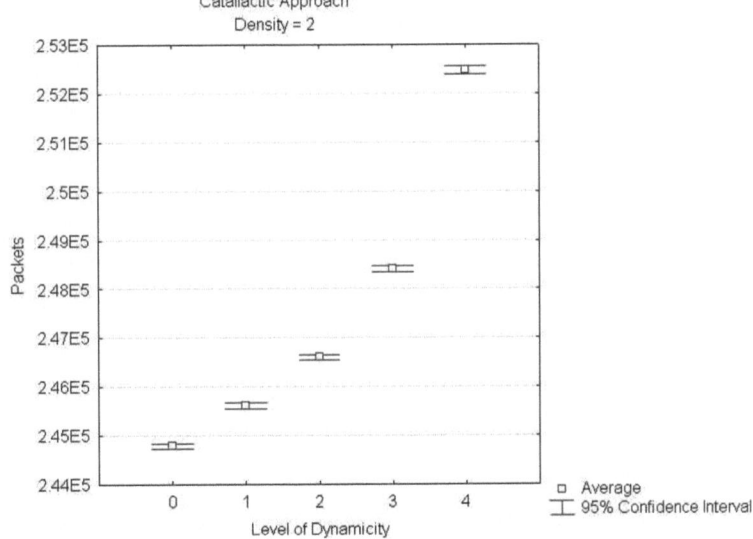

Abb. 10: *Mittlere Anzahl der Pakete und 95%-Konfidenzintervall ohne Resource Broker*

3.3.4 Synopse

Zusammenfassend zeigt sich die Robustheit des dezentralen Ansatzes bei variierenden Eingangsparametern: Während die zentrale Ressourcenallokation mit einem Resource Broker Unzulänglichkeiten in Bezug auf Skalierbarkeit aufweist und eine Unfähigkeit im Umgang mit Dynamik im Netzwerk wie Verbindungs- und Knotenausfälle zeigt, reagieren katallaktische, dezentrale Ansätze besser auf diese Umstände. Steigende Dynamik kann durch ein Erhöhen der Dichte ausgeglichen werden, was einer breiteren physikalischen Verteilung von Serviceinstanzen gleich gesetzt werden kann. Weiterhin deuten die Ergebnisse darauf hin, dass Resource Broker Skalierbarkeitsprobleme haben, die in dezentralen Systemen nicht zu beobachten waren. Diese Robustheit bringt allerdings Kosten mit sich, so steigt die Anzahl der benötigten Nachrichten deutlich an.

3.4 Fazit und Ausblick

Der vorliegende Beitrag bezieht sich auf Dienstauswahlverfahren in serviceorientierten Overlaynetzen. Die derzeit verwendeten Verfahren für Dienstselektion bzw. Dienstauswahl im On-demand Computing weisen Mängel und Unvollständigkeiten auf, teilweise in Bezug auf Skalierfähigkeit und in der Qualität der Allokation, so dass sie nicht uneingeschränkt zu empfehlen sind.

Mit der vorgestellten und durchgeführten Simulation konnte exemplarisch gezeigt werden, dass die Dienstauswahl zunächst überhaupt einen Einfluss auf die Leistungsfähigkeit hat und weiterhin, dass dezentrale Dienstauswahlmethoden unter Ausschluss eines Resource Brokers mit ökonomisch motivierten Verfahren in verteilten, dynamischen Netzen im Vergleich bessere Ergebnisse erreichen. Die Robustheit gegenüber wachsender Ausfallwahrscheinlichkeit und breiter Verteilung der Dienste ist deutlich erkennbar. Der prototypische Konzeptbeweis in einer existierenden, heterogenen, serviceorientierten Infrastruktur wird derzeit innerhalb des EU-Projektes CATNETS erarbeitet [18].

Themen wie „IT-Utility", „Adaptive Computing" oder „Utility Computing", also der beliebige Bezug von IT-Dienstleistungen je nach Bedarf werden in der Wirtschaftsinformatik als ein neues Paradigma diskutiert. Für Unternehmen werden diese Begriffe aber erst dann eine realistische Option zur Eigenbereitstellung von Diensten, wenn sich das Thema On-demand Computing so weit etabliert hat, dass die nötigen Kapazitäten zur Verfügung stehen und die geforderten architektonischen Eigenschaften erfolgreich umgesetzt sind.

Die Frage, ob und bis wann on-demand ein etabliertes Phänomen wird, bleibt. Die IBM will in den nächsten Jahren bis zu 10 Milliarden USD jährlich ausgeben, um ODC erfolgreich auf dem Markt zu platzieren [47]. Ob es sich dabei um ein nachhaltiges Modell handelt, wird schließlich abhängig sein von einer großen Anzahl von Einflussfaktoren, einschließlich des breiten Einsatzes der Kerntechnologien, um die Initiative in den nächsten Jahren unterstützen. Zusätzlich müssen IT-Manager die Herausforderung annehmen, diesen kulturellen Wandel zu führen [48]. Es bleibt außerdem abzuwarten, ob die Rechenleistung verschiedener Anbieter tatsächlich frei austauschbar wird, so wie es bei den ganzheitlichen Ansätzen Grid Computing und On-demand Computing geplant ist. Die dunkle Alternative dazu wären Grids, in denen offene Standards ersetzt würden, um Kunden durch hohe Wechselkosten in proprietärer Technologie einzuschließen.

3.5 Literaturverzeichnis

[1] On-demand computing. http://searchcio.techtarget.com/sDefinition/0,,sid19_gci903730,00.html, Abruf am 2003-11-10.

[2] Bloomberg, J.: Just what is "on demand," anyway? http://searchwebservices.techtarget.com/tip/0,289483,sid26_gci900710,00.html, Abruf am 10.11.2003. {nur für geschlossene Benutzergruppe zugänglich}

[3] Frank, M.: On-Demand-Computing ist eine Managementaufgabe. http://www.computerzeitung.de/O/50/Y/84021/VI/10061386/default.aspx, Abruf am 2003-11-11.

[4] Rechenkraft.net Homepage. http://www.rechenkraft.de, Abruf am 2004-07-26.

[5] Eggs, H.: Vertrauen im Electronic Commerce – Herausforderungen und Lösungsansätze. Deutscher Universitätsverlag, Wiesbaden 2001, S.249.

[6] Picot, A.; Reichwald, R.; Wigand, R.: Die grenzenlose Unternehmung. Information, Organisation und Management. Gabler, Wiesbaden 1996, S.41.

[7] Service-oriented Architecture (SOA) definition. http://www.service-architecture.com/web-services/articles/service-oriented_architecture_soa_definition.html, Abruf am 2004-06-23.

[8] W3C: Web Services Description Language (WSDL) 1.1. http://www.w3.org/ TR/wsdl, 2001-03-15, Abruf am 2004-06-27.

[9] Webopedia: What is UDDI? http://www.webopedia.com/TERM/U/UDDI.html, Abruf am 2004-06-27.

[10] SOAP Version 1.2 Part 1: Messaging Framework. http://www.w3.org/ TR/2003/REC-soap12-part1-20030624/, Abruf am 2004-10-12.

[11] Sabbah, D.: Bringing Web Services Together. http://www.globus.org/wsrf/sabbah_wsrf.pdf, Abruf am 2004-06-25.

[12] Krill, P.: IBM proposes convergence of Web services, grid computing. http://www.infoworld.com/article/04/01/20/HNgridspecs_1.html, 2004-01-20, Abruf am 2004-07-27.

[13] Korupolu, M.R.; Dahlin, M.: Coordinated Placement and Replacement for Large-Scale Distributed Caches. Proc. of IEEE Workshop on Internet Applications, 1999.

[14] Rabinovich, M.; Aggarwal, A.; RaDaR: A scalable architecture for a global Web hosting service. The 8th International World Wide Web Conference, Toronto 1999.

[15] Akamai Webseite, 2004, http://www.akamai.com, Abruf am 200405-21.

[16] Clark, I.: A distributed decentralised information storage and retrieval system. 1999, http://freenet.sourceforge.net/Freenet.ps. Abruf am 2002-12-8.

[17] Foster, I., Kesselman, C.: Globus: A Metacomputing Infrastructure Toolkit. International Journal of Supercomputing Applications 11(2) (1997), S. 115-129.

[18] CATNETS Projektantrag. Nichtveröffentlichter Antrag an die europäische Kommission, 2003b.

[19] Milojicic, D.S.; Kalogeraki, V.; Lukose, R.; Nagaraja, K.; Pruyne, J.; Richard, B.; Rollins, S.; Xu, Z.: Peer-to-Peer Computing. Report No. HPL-2002-57. Palo Alto: Hewlett Packard

3.5 Literaturverzeichnis

Labs 2002, http://www. hpl.hp.com/techreports/2002/HPL-2002-57.html, Abruf am 2002-05-01.

[20] Snelling, D.; Priol, T. et al.: Next Generation Grid(s). European Grid Research 2005 - 2010. Brussels: Information Society – DG, Grids for Complex Problem Solving 2003, http://www.cordis.lu/ist/grids/index.htm, Abruf 2004-06-25.

[21] Ardaiz, O.; Freitag, F..; Navarro, L.: Multicast Injection for Application Network Deployment. 26nd IEEE Conference on Local Computer Networks. Tampa, USA, 2001.

[22] Bell, W.; Cameron, D.; Capozza, L.; Millar, P.; Stockinger, K.; Zini, Floriano: OptorSim - A Grid Simulator for Studying Dynamic Data Replication Strategies. International Journal of High Performance Computing Applications, 17(4), 2003.

[23] Silberstein, M.; Factor, M.; Lorenz, D.: DYNAMO - DirectorY, Net Archiver and Mover. In: Proceedings of Grid Computing - GRID 2002: Third International Workshop, Baltimore, MD, USA, 18. November, 2002. Lecture Notes in Computer Science 2536, Springer-Verlag Heidelberg, 2003: S. 256 – 267.

[24] Ratnasamy, S.; Francis, P.; Handley, M.; Karp, R.; Shenker, S.: A Scalable Content-Addressable Network. http://citeseer.nj.nec.com/ratnasamy01scalable.html, Abruf am 2004-06-20.

[25] Balakrishnan, H.; Kaashoek, F.; Karger, D.; Morris, R.; Stoica, I.: Looking up data in P2P systems, Communications of the ACM, February 2003, S. 43-48.

[26] Frey, J.; Tannenbaum, T.; Livny, M.; Foster, I.T.; Tuecke, S.: Condor-G: A Computation Management Agent for Multi-Institutional Grids. Cluster Computing 5(3) (2002), S. 237-246.

[27] Litzkow, M. J.; Livny, M.; Mutka, M.: Condor – a hunter of idle workstations. In: Proceedings of the 8th International Conference of Distributed Computing Systems, San Jose, Kalifornien, 1988.

[28] Liu, C.; Yang, L.; Foster, I.; Angulo, D.: Design and Evaluation of a Resource Selection Framework for Grid Applications. In: Proceedings of the 11 th IEEE International Sym-

posium on High Performance Distributed Computing HPDC-11 2002 (HPDC' 02) (2002), S. 63.

[29] Chandra, P.; Fischer, A.; Kosak, C.; Ng, E.; Steenkiste, P.; Takaha-shi, E.; Zhang, H.: Darwin: Customizable Resource Management for Value-Added Network Services. In: Proc. of the Sixth IEEE International Conference on Network Protocols (ICNP' 98), Austin/Texas, 1998.

[30] Bestavros, A.: Demand-based Dissemination for Distributed Multi-media Application. In: Proceedings of the ACM/ISMM/IASTED International Conference on Distributed Multimedia Systems and Applications, Stanford, Kalifornien, 1995.

[31] Amir, E.; McCanne, S.; Katz, R.H.: An active service framework and its application to real-time multimedia transcoding. In: Proceedings of ACM SIGCOMM' 98, Vancouver, Canada, 1998.

[32] Ardaiz, O.; Artigas, P.; Freitag, F.; Navarro, L.; Eymann, T.; Reinicke, M.: Decentralized Resource Allocation in Application Layer Networks. In: Proceedings of the 3rd IEEE/ACM International Symposium on Cluster Computing and the Grid (CCGrid 2003), 12.-15. Mai 2003, Tokyo, Japan.

[33] Adar, E.; Huberman, B.A.: Free Riding on Gnutella. First Monday 5(10) (2000).

[34] Buyya, R.: Economic-based Distributed Resource Management and Scheduling for Grid Computing. Ph.D. Thesis. Monash University, Melbourne, Australia, 2002, http://www.buyya.com/thesis/thesis.pdf, Abruf 2004-06-19.

[35] Tesfatsion, L.: How economists can get alife. In: Arthur, W.B., Durlauf, S., Lane, D.A. (eds.): The Economy as a Evolving Complex System II. Santa Fe Institute Studies. Redwood City, CA: Addison Wesley 1997: S. 533-564.

[36] Press, W. H.; Teukolsky, S. A.; Vetterling, W. T.; Flannery, B. P.: Numerical Recipes in C++ - The Art of Scientific Computing. Cambridge University Press: Cambridge, Massachusetts, 2002.

[37] Brenner, T.: Learning in a Repeated Decision Process: A Variation-Imitation-Decision Model. Report No. #9603. Max-Planck-Institut für die Erforschung von Wirtschaftssystemen: Jena 1996.

3.5 Literaturverzeichnis

[38] Kagel, J.H., Roth, A.E.: The handbook of experimental economics. Princeton, N.J: Princeton University Press 1995.

[39] Pruitt, D.G.: Negotiation behavior. In: Organizational and occupational psychology. New York: Academic Press 1981

[40] Smith, V.L.: An experimental study of competitive market behavior. In: Journal of Political Economy, Vol. 70 (1962): S. 111-137.

[41] Eymann, T.: Avalanche – ein agentenbasierter dezentraler Koordinationsmechanismus für elektronische Märkte. Ph.D. Thesis. Albert-Ludwigs-Universität Freiburg, 2000, http://www.freidok.uni-freiburg.de/ volltexte/147/, Abruf am 2004-05-01.

[42] Miller, M.; Drexler, E.: Markets and Computation: Agoric Open Systems. In: The Ecology of Computation, Bernardo Huberman (Hrsg.) Elsevier/North-Holland 1988.

[43] Cave, D.: The Mojo solution, (9.10.2000), http://archive.salon.com/tech/view/2000/10/09/mojo_nation, erstellt am 2000-11-09.

[44] Mojo Nation Webseite, 2003, http://www.mojonation.net, Abruf am 2003-07-21.

[45] CATNET Project: Catallaxy Evaluation Report. Report No. D3. Barcelona: 2003a, http://research.ac.upc.es/catnet/pubs/D3.pdf, Abruf am 2003-07-24

[46] J-Sim Website. http://www.j-sim.org. Abruf am 2003-02-21.

[47] Farber, D.; On-demand computing: What are the odds?. http://techupdate.zdnet.com/techupdate/stories/main/0,14179,2896789,00.html, erstellt am 2002-06-11.

[48] Shankland, S.: IBM: On-demand computing has arrived. http://news.com.com/IBM:+Ondemand+computing+has+arrived/2100-7784_3-5106577.html?part=business2-cnet, Abruf am 2004-06-26

4 Grid Economics: Market Mechanisms for Grid Markets

D. Neumann, D. Veit und C. Weinhardt

This chapter describes a market-based approach to conduct resource allocation problems in computational Grids. Currently, there is a lot of research done on the technical layer of resource allocation. However, the economic allocation of resources is still an open issue in research and, more importantly, in practical application.

The focus in this chapter is on existing market mechanisms, which are transferred from different domains. Possible applications in Grid scenarios are discussed. Most promising – but also most demanding – are combinatorial auctions and exchanges that include high bidding complexity. It is demonstrated how such mechanisms can be applied to Grid markets and what obstacles have to be overcome in order to successfully deploy these mechanisms in future resource sharing markets.

4.1 Introduction

Grid Computing makes use of computing resources across a distributed and heterogeneous set of computers linked by a network. The Grid denotes the hardware and software infrastructure that provides access to computational resources. These resources can be comprised of CPU cycles, storage, network bandwidth,

4.1 Introduction

and even applications as services. The idea of Grid is to make computing resources on demand available as utilities. Examples for Grid applications include SETI@home, which filter data from particle accelerator or Folding@home, which studies protein folding for discovering antidotes for diseases.

Since the last years, Grid technology has also been gaining attractiveness for commercial organizations, which strive for extremely high computing power for limited time, but are not necessarily willing to invest further in their own computing resource infrastructure. This increasing attractiveness can be document by the large-scale efforts of several big companies such as BMW, Ericsson, IBM, Hewlett-Packard and Unilever in Grid initiatives to keep apace with the ongoing advances in Grid technology and its corresponding business opportunities.

Despite these breathtaking advances in Grid technology, the full potential of Grid is hardly even touched. From a theoretical standpoint, Grid users could utilize resources from multiple locations (e.g. CPU cycles, memory capacities, network bandwidth, or storage media). Typically, Grid users avoid reaching far beyond their home institutions, even though their own resources are often inadequate to conduct computationally demanding applications. This seemingly contradiction is reasoned by the significant barriers that arise when organizational boundaries are crossed. These barriers mainly reflect different mechanisms and policies of different organizational units. In addition to these barriers, resource owners have little incentive to share their resources as sharing is oftentimes not monetarily rewarded. Free-riding as in Peer-to-Peer systems hence becomes a factor, reducing the overall success of Grid.

Obviously, those problems – impeding the wide adoption of Grid – affect both, the resource users and owners. In essence, the problems occur because traditional resource scheduling mechanisms do not provide the right incentives for rewarding resource sharing and for demanding resources over the Grid.

The use of market mechanisms for scheduling resources is often motivated as they provide the right incentives for participating: By the interplay of demand and supply, competition can achieve an efficient allocation in a way that those jobs are awarded with resources, which are most valuable. Since the resource owners receive real money (instead of funny money [1]), it is expected that idle resources are supplied on the Grid market.

This work is structured as follows: In section 2 the Grid environment is characterized and the associated requirements are defined. Section 3 discusses market mechanisms with respect to their applicability in Grid. As a result, a class of mechanisms is identified, which is appropriate for the Grid. Section 4 addresses the impediments, which may explain why market mechanisms have not made it into practice yet. Section 5 concludes with a summary.

4.2 The Grid Environment and Corresponding Requirements

A market mechanism consists of rules, which govern how bids for grid can be formulated and how they are translated into an outcome (i.e. allocation and corresponding prices). In other words the market mechanism defines the incentive structure of the market. The incentive structure, however, is not fixed, but depends on the underlying economic environment. The economic environment denotes all factors that have an impact on demand and supply, but is independent on the market mechanism. Those factors accordingly embrace information about the potential market participants, their needs, the characteristics of the traded resources, and the participants' endowments. This simple relationship between the market mechanism, the underlying environment and the outcome has major ramifications on the design of market mechanisms. When proposing a new market mechanism, it is essential that the market mechanism solves the allocation problem that the environment exposes. Stated differently, an adequate design of market mechanisms commences with a thorough understanding of the potential market participants, their drivers and motives, and the economic problem that the market is trying to solve [2].

Following this intuition, this section analyzes the economic environment of Grid computing, which raises many requirements the potential market mechanism has to meet.

4.2.1 The Grid Environment

Surveying the grid environment and following [3], the simplified picture in Fig. 1 can be drawn for Grid resources. In essence, the participants of the Grid market are resource owners as sellers

4.2 The Grid Environment and Corresponding Requirements

(e.g. IBM or Sun with their computer centers), the resource consumers as buyers (e.g. scientists at universities, rendering or the biochemical firms) and some intermediaries (e.g. Condor, Gallop, Legion etc.) that technically provide the resource management infrastructure for exploiting remote resources. According to the resource management architecture proposed by [3] the intermediary layer consists of three basic components:

- Resource Broker: The resource broker components are responsible for resource discovery, selection, aggregation, and subsequently for the data and program transportation [4]. By transforming the resources to the consumers' requirements (which are specified for instance in the Resource Specification Language (RSL)) into a set of jobs that are self/reliantly scheduled on the appropriate resources (i.e. RSL specialization) and subsequently managed, the complexity of the Grid is concealed. For the market participants the resource broker is apparently more of a black box.

- Resource Information Manager: The resource information manager provides pervasive access to information about the current availability and capability of resources [3].

- Allocator: The allocator coordinates the allocation of resources at multiple sites. Obviously, the allocator and the information service assume the responsibility of (meta-) scheduling the jobs.

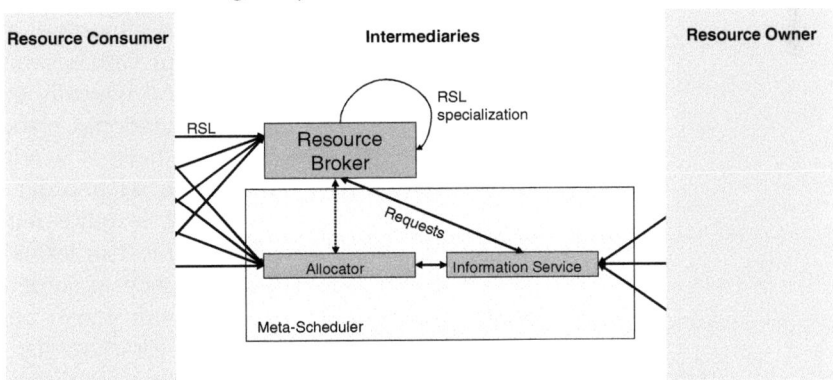

Fig. 1: Structure of the Grid Market

Based upon this view on the intermediary layer, the market mechanism for Computational Grid can be sketched as follows:

4 Grid Economics: Market Mechanisms for Grid Markets

The transition from an intermediary layer to a mediated market mechanism is not too far. In essence, the scheduling performed by the resource broker can be shifted to the market mechanism. Instead of sending requests to the information service, the resource broker can translate the user requirements into bids. Those bids expressing demand and supply situation are subsequently cleared by the market mechanisms.

4.2.2 Requirements

Following the principles of Market Engineering [5], which define a prescriptive approach for designing and analyzing electronic markets, it is necessary that the mechanism accounts for the requirements on the outcome (what is to achieve?) and on the mechanism itself (how it is achieved). While the requirements on the outcome reflect well-known economic measures, the requirements on the mechanism account for the domain related requirements, i.e. the Computational Grid.

4.2.2.1 Requirements on the outcome

In literature, the analysis of the outcome of a mechanism comprises the following standard measures [6, 7]:

- Allocative Efficiency: If utility is transferable among all participants, a mechanism that maximizes the sum of individual utilities (i.e. the sum of surpluses conditional on the given information set) is called allocative efficient. Allocative efficiency is particularly for Grid resources a very important objective. Since demand typically exceeds supply, there are not sufficient computational resources available to compute all jobs. Hence, there is a criterion needed, which states what jobs and in what order are computed by the available resources. It is quite natural to award those jobs, which are valued most in terms of utility. And this refers to allocative efficiency as super-ordinate concept. Nonetheless, when users have complex needs, achieving this goal is not easy for either the participants or the market mechanism [8].

- Incentive Compatibility: Incentive compatibility refers to the validity of messages the participants place. A mechanism is incentive compatible if the participants report their preferences truthfully as a dominant strategy. Thus, par-

4.2.2 Requirements

ticipants may not have an incentive to untruthfully report their preferences in order to increase their individual utility.

- Individual Rationality: The constraint of individual rationality requires that the utility after participating in the mechanism must be greater or equal than before. Otherwise the participants would decide not to take part in the mechanism.

- Budget Balance: A mechanism is said to be strictly budget balanced if the amount of prices sum up to 0 over all participants. In this case the mechanism redistributes the payments among the sellers and buyers. Neither are funds removed from the system nor is the system subsidized from outside. Strict budget balance is an important property since the resource allocation can be performed at no costs. In case the mechanism runs a deficit, it has to be subsidized by some outside source and is thus not per-se feasible [9, 6].

- Computational Tractability: Computational tractability considers the complexity of computing the outcome of a mechanism from the agents' strategies. With an increasing size of messages, the allocation problem can become very demanding. Thus, computational constraints may delimit the design of choice and transfer rules [10, 11]. Due to the complexity of the allocation problem, it may even happen that the previously mentioned allocative efficiency and incentive compatibility requirements are totally relaxed.

Generally, the allocative efficiency qualifies for objective functions the mechanism designer wants to achieve. The remaining categories are constraints upon the objective function [12].

4.2.2.2 Requirements upon the mechanism

To achieve the above mentioned outcome requirements, an adequate mechanism has to be designed. As aforementioned a mechanism consists of a bidding language, a winner determination and pricing rule. With respect to the Computational Grid environment, the following requirements can be elicited upon the mechanism [13]:

4 Grid Economics: Market Mechanisms for Grid Markets

- Double-sided mechanism: The mechanism apt for Computational Grid must enable many resource owners (henceforth sellers) and many resource consumers (buyers) to trade simultaneously.

- Language includes combinatorial bids: Buyers usually demand a combination of computer resources as a bundle to perform a task [14]. As such, computer resources in the Grid are complementarities, i.e. participants have super-additive valuations for the resources, as the sum of the valuations for the single resources is less than the valuation for the whole bundle ($v(A)+v(B) \le v(AB)$). Suppose a buyer requires hard disk space, a CPU, and memory to render images. If any component, e.g. the CPU, is not allocated to him, the remaining bundle has no value for him since the rendering job cannot be processed without the CPU. In order to avoid this exposure risk (i.e. receiving only a subset of the bundle), the mechanism must allow for bids on bundles. Likewise, the seller must also be able to express bids on bundles.

  The buyer may want to submit more than one bid on a bundle but many that are excluding each other. In this case, the resources of the bundles are substitutes. This means that the buyer has sub-additive valuations for the resources ($v(A) + v(B) \ge v(AB)$). For instance, a buyer is willing to pay a high price for a job during the day and a low price if the job is executed at night. However, this job must be computed only once. As such, the mechanism must support XOR-bids to express substitutes.

  For simplicity it is also possible that the bids of the sellers are reduced to a set of OR-bids. This simplification can be reasoned by the fact that resources in the Computational Grid are non-storable commodities, i.e. the CPU power available at a present time cannot be stored for a later time.

- Language includes bids on attributes: Resources in Grids are typically not completely standardized as similar resources can differ in their quality. For instance, a hard disk can be characterized by the following quality attributes: capacity (in Gigabyte GB), access time (in milliseconds (ms)), and data throughput (in bits per second (bits/s)). Thus, the mechanism is required to support bids on multiple attributes whereas it is assumed that a job can be conducted, when at least the minimum quality requirements of similar resources are met (this concept can also be extended to the use of generalized policy descriptions [15]). For example, while a rendering job requiring a minimum amount of GB, say 250 GB, can be conducted by a 500 GB hard disk, it cannot by a 100 GB hard disk.

- Language includes bids on time-slots: Buyers usually require resources only for a specific time span. Having conducted the computation, there is typically no further use for the resources. The exact timing of the computation is not always that important for the buyer. For instance, the buyer may be indifferent whether the job is performed at 10 a.m. or at 11 a.m., as long as the job is finished at certain time, e.g. 3 p.m. Therefore, the mechanism must allow for placing bids on time ranges.

- Clearing and pricing rules that exploit the full-range of the language. Furthermore, clearing and pricing rules have to be designed that (1) impute a desirable allocation (allocative efficient) and (2) make usage of all information of the bidding language.

4.3 Market Mechanisms for Grid

A number of mechanisms have been proposed that attempt to solve the resource allocation problem in such a Grid market or related architectures. Most of these mechanisms are central in nature in a way that the allocation problem is solved by a central entity using global optimization algorithms without the employment of prices. This central entity requires detailed information about the demand and supply situation in order to be effective. As information is dispersed among the buyers and sellers, central allocation algorithms may not enfold their full power, because this information requirement is not even closely met. Market-based approaches incorporate incentives for truthful information revelation by implementing prices. Surveying the literature, various mechanisms for allocating computer resources adapt classical auction mechanism for standardized products as described in section 3.1. Furthermore, there exists a number of multi-attribute, bundle, or combinatorial mechanisms, which can be used to trade dependent and non-standardized products (section 3.2). To be applicable for allocating computational resources, the market mechanisms must also account for time attributes. Section 3.3 surveys combinatorial auctions and exchanges that incorporate a scheduling problem. Market mechanism are per-se no panacea for allocating resources. All mechanisms have advantages and disadvantages. In many cases, market mechanisms are proposed as adequate resource allocation mechanism for computational grid without reviewing the properties of the mechanisms. In the following, all three mechanism types are discussed with respect to their applicability in the Grid market [16].

4.3.1 Classic Auction Types

There exist a number of economic resource allocation systems which adapt classical auction mechanisms (e.g. English auctions) for Peer-to-Peer networks and Grid Computing infrastructures.

- Standard Vickrey Auction: The SPAWN system provides a market mechanism for trading CPU times in a network of workstations [17]. SPAWN treats computer resources as standardized commodities and implements a standard Vickrey auction. It is known from auction theory that the Vickrey auction attains (1) truthful preference revelation and (2) an efficient allocation of resources [18]. However, SPAWN does not make use of the Generalized Vickrey auction, which can cope with super additive valuations. Furthermore, the Vickrey auction can neither cope with multiple attributes nor with different time slots. Hence, SPAWN cannot meet the aforementioned requirements.

- Repeated Vickrey Auction: POPCORN provides an infrastructure for global distributed computation [19, 20]. POPCORN mainly consists of three entities: (i) A parallel program which requires CPU time (buyer), (ii) a CPU seller, and (iii) a market which serves as meeting place and matchmaker for the buyers and sellers. Buyers of CPU time can bid for one single commodity, which can be traded executing a Vickrey auction repeatedly. POPCORN obviously fails with respect to the same shortcomings as SPAWN.

- Distributed Sealed-bid Continuous Double Auction: OCEAN (Open Computation Exchange and Arbitration

4.3.1 Classic Auction Types

Network) is a market-based infrastructure for high-performance computation, such as Cluster and Grid computing environments [21, 22]. The major components of the OCEAN's market infrastructure are user components, computational resources, and the underlying market mechanism (e.g. the OCEAN Auction Component). In the OCEAN framework, each user (i.e. resource provider or consumer) is represented by a local OCEAN node. The OCEAN node implements the core components of the system, for instance a Trader Component, an Auction Component, or a Security Component. The implemented OCEAN auctions occur in a distributed Peer-to-Peer manner. The auction mechanism implemented in the OCEAN framework can be interpreted as a distributed sealed-bid continuous double-auction [21]. A trade is proposed to the highest bidder and the lowest seller. Afterwards, the trading partner can renegotiate their service level agreements. The renegotiation possibility one the one hand allows to cope with multiple attributes and with the assignment of resources to time slots. Nonetheless, the negotiation makes the results of the transparent auction obsolete. Neither the auction can enfold its full potential, nor can the negotiation guarantee to achieve an efficient allocation, as competition is trimmed.

- Commodity Markets: G-Commerce provides a framework for trading computer resources (CPU and hard disk) in commodity markets and Vickrey auctions [23, 24, 25]. While the Vickrey auction has the abovementioned shortcomings in grid, the commodity market typically works with standardized products. Additionally, the commodity market cannot account for the complementarities among the resources, as only one leg of the bundle is auctioned off, exposing the bidder to the threshold risk.

- Cooperative Bartering and Sharing: The Stanford Peers model is a Peer-to-Peer system which implements auctions within a cooperative bartering model in a cooperative sharing environment [26]. It simulates storage trading for content replication and archiving. It demonstrates distributed resource trading policies based on auctions by simulation.

Reviewing the requirements on the mechanism, it becomes evident that the previous described mechanisms fail to satisfy the requirements of the Grid market. Especially the negligence of time attributes for bundles and quality constraints for single resources diminish the use of the proposed market mechanisms.

4.3.2 Combinatorial Auctions and Exchanges

Combinatorial Auctions and Exchanges comprise dependencies between resources and have attained a lot of attention in the last few years [9, 27, 28, 29, 30, 13]. Combinatorial Auctions and Exchanges allow submitting logical concatenated bids like AND bids, OR bids, and XOR bids. The use of these auction types is crucial if resources are complementarities or substitutes. Due to their capability of guaranteeing "all-or-nothing"-bids, their application for trading computer resources is deemed promising. At the moment, there exist different combinatorial and bundle mechanisms:

- Generalized Vickrey Auction: The Generalized Vickrey Auction (GVA) is a Vickrey-Clarke-Groves mechanisms (cf. Vickrey Auction) and implements a single shot and sealed bid second-price approach for combinatrial auctions. As the single item Vickrey auction, the GVA is strategy-proof, i.e. truthful bidding is a dominant strategy [31]. However, the mechanism is single sided and, thus, does not generate competition on both sides. Even worse, the GVA treats the traded resources as standardized commodities and does not support quality characteristics. Additionally, the computational costs of the GVA are considerably high, as it requires solving several instances of a NP-hard problem [9].

- Dutch Combinatorial Auctions: In a Dutch Combinatorial Auction, the auctioneer calls out a price for each bundle and lowers this price incrementally as long as no bidder is willing to accept it. An implementation of a Dutch Combinatorial Auction can be found in [28]. However, the auction is single-sided and does neither support quality characteristics nor time attributes of resources.

- Iterative Combinatorial Auctions: Iterative Combinatorial Auctions adapt English Auctions for the combinatorial case

[32, 33]. Iterative combinatorial auctions are the most prominent combinatorial mechanisms. Nonetheless, it is not possible to make us of them in the Grid market, as they do not account for multiple attributes.

- Bundle Exchanges: Bundle exchanges are double auctions with support AND concatenated bids on bundles. In contrast to combinatorial auctions and exchanges, these auctions types neglect the use of XOR bids. An implementation of a bundle exchange can be found in [34]. Classical bundle exchanges neglect the support of quality characteristics, which prevents their direct application in the Grid market. However, if the requirements for the Grid market are incorporated, a bundle exchange may be appropriate for computational resources. Furthermore, the bundle exchange does not support XOR orders making the allocation problem computationally more tractable.

- Combinatorial Exchanges: Combinatorial exchanges are double auctions that support combinatorial orders (XOR orders). Parkes et al. introduce the first combinatorial exchange as a single-shot sealed bid auction [35]. Biswas and Narahari propose an iterative combinatorial exchange based on a primal/dual programming formulation of the allocation problem [28]. However, both approaches neither account for time nor for quality constraints and are, thus, not directly applicable in the Grid market. However, the ability of trading dependent resources simultaneously qualifies for a Combinatorial Exchange that meets most of the requirements for allocating computational resources.

Reviewing the requirements on the mechanism, it becomes obvious that the previously described mechanisms fail to satisfy these requirements. In particular, the negligence of time attributes for bundles and quality constraints for single resources diminish the use of the proposed market mechanisms. The introduction of time attributes redefines the allocation problem to a type of scheduling problem.

4.3.3 Scheduling Auctions

To account for time attributes, Wellman et al. suggest a single-sided auction protocols for the allocation and scheduling of resources under consideration of different time constraints [36].

Conen goes one step further by designing a combinatorial bidding procedure for job scheduling including different running, starting, and ending times of jobs on a processing machine [37]. Bapna et al. develop a combinatorial call auction that also incorporates time attributes. They present three pricing schedules that trade off the economic properties of allocative efficiency, incentive compatibility, and fairness in allocation with computational costs. The first pricing scheme draws on Vickrey-Clarke-Groves prices. The welfare maximizing allocation is achieved, but is computationally demanding, as it requires the solving of multiple instances of an NP-hard problem. The second pricing scheme adds fairness constraints to the pricing scheme, reducing the computational burden to some extent – yet preserving incentive compatibility. The third pricing scheme uses a time sensitive fair Grid heuristics (tsfGRID) that accounts for real-time fast solution techniques. In essence, tsfGRID relaxes the allocative efficiency requirement of the optimal fair solution. Nevertheless, the solution is not guaranteed to be incentive compatible, but the heuristic is designed to be fast and maintain fairness in allocations [38]. All three scheduling auctions are, however, single-sided and thus do not create competition on both sides. Furthermore, co-allocation in a way that several resource owners can satisfy one job request is not possible in those mechanisms.

The MACE (Multi-Attribute Combinatorial Exchange) mechanism is intended to remedy these shortcomings by providing a combinatorial exchange that accounts for both multi-attributes and time constraints. The installation of competition on both sides guarantees that co-allocation is possible. The suggested k-pricing scheme distributes the surplus of a transaction (difference between the prices of two corresponding buy and sell bids) among the participants. As such, the k-pricing is budget-balanced, i.e. does not require subsidies from outside the mechanism as this is the case with Vickrey-Clarke-Groves prices, and is comparably fast. The drawback of the MACE mechanism is that the incentive compatibility and allocative efficiency properties are not achieved. However, stochastic simulations have revealed that both properties are at least *approximately* attained [39, 40]. The MACE mechanism is, however, also hampered by the complexity of the allocation problem.

Comprising, the demonstrated scheduling auctions are applicable in the Grid market. Interestingly, in literature suggesting market mechanisms for resource allocation, it is typically referred to classical auction types, which are inapplicable for that environment, while more complex auctions are step-motherly treated.

4.4 Impediments of Market Mechanism Adoption

Obviously, there are market mechanisms available that can be used for allocation computational resources. But despite that, an active marketplace for trading computational resources does not yet exist. There are several impediments hindering the adoption of market places. In essence, the market mechanisms require that buyers and sellers submit bids specifying their demand and supply. This, however, suggests that buyers and sellers know their exact demand and supply and, moreover, can assign a value to their bid. In addition, the assumptions that any job can be computed everywhere in the Grid is rather unrealistic. Instead, there may be highly sensitive jobs that should only be worked off inside the organization or inside a consortium. The main impediments of market mechanism adoption can be summarized as follows:

- Accounting: In order to match enough buyers with sellers, current market-based resource allocation schemes batch allocations into blocks of time. The time scale of this batch system can be minutes or days ahead of when the resources will actually be made available. This means that users must predict their resource needs in advance. It is difficult to precisely describe the level of resources required to run an experiment or job. Depending on the inputs to a program, the ideal level of resource consumption can vary considerably. This has some major ramifications on the bidding process, since market mechanisms can achieve a desirable efficient allocation of resources by aggregating dispersed information, if buyers and sellers know their demand and supply situation quite accurately. If buyers and sellers have not this information, underbidding as well as overbidding may occur, which is both undesirable.

  A resource underbid will prove unsatisfying if won as the job cannot be conducted, while a resource overbid (with the same value) is less likely to win because of competition from more efficient users. Requiring users to predict their resource need can be difficult and can be an obstacle for the installation of market mechanisms. To overcome the accounting problems, tools are need that to help users estimate their resource needs. Such tools are an open area for ongoing and future research.

- Explication of Policies: The submission of binding offers for services (aggregating computational resources) is a highly critical process, as it addresses many political issues. For instance, companies will neither offer nor demand resources from any arbitrary corporation. Obviously policies are needed that govern strict constraints or preferences on the resource demand or supply. This becomes even more important when the boundaries of organizational units are crossed. Currently, economic research on the institution of policies is missing as well as the corresponding semantic formulization of the policies.

- Pricing Guidelines: Utility maximizing market mechanisms are only as accurate as the prices that users attach to their bids on resources that they own, and resources that they would like to acquire. However, what is the accurate valuation of three hours of CPU time, a week before a major simulation will take place? As before, any situation of excess demand will lead to unsatisfied users [41]. Ultimately, there is a need for pricing guidelines that must shed light into the derivation of prices. Hence, these guidelines formulize the business model of resource providers and the demand function of resource demand.

- Virtualization: Virtualization plays a major role in all Grid-related considerations. It is deemed important that the distribution of computations is conducted transparently for the user. But this also requires that the market process occurs automatically in a self-organizing way. This, in turn, requires the development of tools extending the Grid platform that automatically assess the demand and/or supply situation, translate them into bids, determine the associated prices, and submits those fully-specified bids to the market mechanism.

4.5 Concluding Remarks

The use of Grid, however, is so far often restricted to single applications or to the boundaries of organizations. Offering computer resources over organizational borders is because of political and economic concerns very difficult.

Obviously, the enormous potential of Grid infrastructures and applications to

- enable on-demand computing of highly resource consuming tasks, and to
- improve the utilization of companies' resources

is currently widely left unused.

One way to accelerate the use of the Grid is the establishment of markets. Markets provide an easy way to allocate computer resources quite efficiently even beyond organizational boundaries by means of the price mechanism. Implicitly, markets can thereby resolve the inherent problem of excess demand for Grid resources. The idea of markets in Grid is certainly not new, but none of previous efforts have attained the desired results of a functioning Grid market: One reason for the failures of Grid markets is simple. Since the emphasis of these markets was purely on the market mechanism, it was not paid any attention to the question, how can real companies take part in the market. Stated differently from a Grid Economics perspective, how can companies formulate offers to buy or sell resources with ease? Obviously, market mechanisms are helpful for allocating resources, but they will only enfold their potential if the impediments associated with mechanisms are solved.

4.6 References

[1] Cheliotis, G., C. Kenyon and R. Buyya (2003). Grid Economics: 10 Lessons from Finance. Joint Technical Report, GRIDS-TR-2003-3, IBM Research Zurich.

[2] Cramton, P. (2003). "Electricity Market Design: The Good, the Bad, and the Ugly." Proceedings of the Hawaii International Conference on System Sciences.

[3] Czajkowski, K., I. Foster and C. Kesselman (2004). Resource and Service Management. The Grid 2 - Blueprint for a New Computing Infrastructure, Elsevier. 2: 259–283.

[4] Chetty, M. and R. Buyya (2002). "Weaving computational grids: How analogous are they with electrical grids?" Computing - Science and Engineering 4(4): 61-71.

[5] Weinhardt, C., C. Holtmann and D. Neumann (2003). "Market Engineering." Wirtschaftsinformatik 45(6): 635-640.

[6] Jackson, M. O. (2002). Mechanism Theory. Encyclopedia of Life Support Systems, UNESCO -online.

[7] Neumann, D. (2004). Market Engineering - A Structured Design Process for Electronic Markets. Fakultät für Wirtschaftswissenschaften. Karlsruhe, Universität Karlsruhe (TH).

[8] Shneidman, J., C. Ng, D. C. Parkes, A. AuYoung, A. C. Snoeren, A. Vahdat and B. Chun (2005b). "Why Markets Could (But Don't Currently) Solve Resource Allocation Problems in Systems." Working Paper.

[9] Parkes, D. C. (2001). Iterative Combinatorial Auctions: Achieving Economic and Computational Efficiency. Philadelphia.

[10] Kalagnanam, J. and D. C. Parkes (2003). Auctions, Bidding and Exchange Design. Supply Chain Analysis in the eBusiness Era. S. D. W. a. Z. M. S. David Simchi-Levi, Kluwer Academic Publishing: forthcoming.

[11] Lehmann, D., R. Mueller and T. Sandholm (2005). The Winner Determination Problem. Combinatorial Auctions. P. Cramton, Y. Shoham and R. Steinberg, MIT Press: Chapter 12.

[12] Krishna, V. and M. Perry (1998). "Efficient Mechanism Design." Working Paper.

[13] Schnizler, B., D. Neumann and C. Weinhardt (2004b). Resource Allocation in Computational Grids - A Market Engineering Approach. WeB 2004, Washington 19-31.

[14] Subramoniam, K., M. Maheswaran and M. Toulouse (2002). Towards a micro-economic model for resource allocation in grid computing systems. IEEE Canadian Conference on Electrical & Computer Engineering.

[15] Lamparter, S., A. Eberhart and D. Oberle (2005). Approximating Service Utility from Policies and Value Function Patterns. International Workshop on Policies for Distributed Systems and Networks, IEEE Computer Society.

[16] Schnizler, B., D. Neumann, D. Veit, M. Napoletano, M. Catalano, M. Gallegati, M. Reinicke, W. Streitberger and T. Eymann (2005a). "D1.1: Environmental Analysis for Application Layer Networks." CATNETS Project Deliverable.

4.6 References

[17] Waldspurger, C. A., T. Hogg, B. A. Huberman, J. O. Kephart and W. S. Stornetta (1992). "Spawn: A distributed computational economy." IEEE Transactions on Software Engineering 18(2): 103–117.

[18] Krishna, V. (2002). Auction Theory. San Diego, CA, Academic Press.

[19] Nisan, N., S. London, O. Regev and N. Camiel (1998). Globally distributed computation over the internet - the popcorn project. 18th International Conference on Distributed Computing Systems, Amsterdam, The Netherlands, IEEE Computer Society.

[20] Regev, O. and N. Nisan (1998). The POPCORN market - an online market for computational resources. First international conference on Information and computation economies, Charleston, South Carolina, ACM Press: 148 - 157.

[21] Acharya, N., C. Chokkaredd, P. Desai, R. Devrajan, M. P. Frank, S. N. Chakravarthul, M. Nagendranath, P. Padala, H. Park, G. Sur, M. Tobias, C.-K. Wong and B. Yu (2001). The open computation exchange & auctioning network (ocean), Department of Computer & Information Science & Engineering, University of Florida.

[22] Padala, P., C. Harrison, N. Pelfort, Erwin Jansen, M. Frank and C. Chokkareddy (2003). Ocean: The open computation exchange and arbitration network, a market approach to meta computing. International Symposium on Parallel and Distributed Computing.

[23] Wolski, R., J. Plank, J. Brevik and T. Bryan (2001b). G-commerce: Market formulations controlling resource allocation on the computational grid. Proceedings of the International Parallel and Distributed Processing Symposium (IPDPS).

[24] Wolski, R., J. Plank, J. Brevik and T. Bryan (2001a). "Analyzing market-based resource allocation strategies for the computational grid." International Journal of High Performance Computing Applications 15(3): 258–281.

[25] Wolski, R., J. Brevik, J. Plank and T. Bryan (2003). Grid resource allocation and control using computational economies. Grid Computing - Making The Global Infrastructure a Reality, John Wiley & Sons: chapter 32.

[26] Cooper, B. F. and H. Garcia-Molina (2002). Bidding for storage space in a peer-to-peer data preservation system. 22nd International Conference on Distributed Computing Systems, IEEE Computer Society: 372.

[27] Sandholm, T., S. Suri, A. Gilpin and D. Levine (2002). Winner Determination in Combinatorial Auction Generalizations. International Joint Conference on Autonomous Agents and Multiagent Systems, Bologna, Italy69-76.

[28] Biswas, S. and Y. Narahari (2003). "Iterative dutch combinatorial auctions." Annals of Mathematics and Artificial Intelligence: forthcoming.

[29] Gradwell, P. (2004). "Distributed combinatorial resource scheduling." Working Paper.

[30] Jain, R. and P. Varaiya (2004). "Combinatorial exchange mechanisms for efficient bandwidth allocation." Communications in Information and Systems 3(4): 305–324.

[31] Green, J. and J. J. Laffont (1977). "Characterization of Satisfactory Mechanisms for the Revelation of Preferences for Public Goods." Econometrica 45(2): 427-438.

[32] Parkes, D. C. (1999). iBundle: An efficient ascending price bundle auction. ACM Conf. on Electronic Commerce, 148-157.

[33] Ausubel, L. and P. R. Milgrom (2002). "Ascending auctions with package bidding." Frontiers of Theo. Economics 1(1): 1-43.

[34] Grunenberg, M., B. Schnizler, D. Veit and C. Weinhardt (2005). Innovative Handelssysteme für Finanzmärkte und das Computational Grid. 67. wissenschaftliche Jahrestagung des Verbandes der Hochschullehrer für Betriebswirtschaft, Kiel.

[35] Parkes, D. C., J. Kalagnanam and M. Eso (2001). Achieving budget-balance with vickrey-based payment schemes in exchanges. International Joint Conference on Artificial Intelligence1161–1168.

[36] Wellman, M. P., W. E. Walsh, P. Wurman and J. MacKie-Mason (2001). "Auction protocols for decentralized scheduling." Games and Economic Behavior 35(271-303).

[37] Conen, W. (2002). Economically coordinated job shop scheduling and decision point bidding - an example for

4.6 References

economic coordination in manufacturing and logistics. Workshop on Planen, Scheduling und Konfigurieren, Entwerfen, Freiburg.

[38] Bapna, R., S. Das, R. Garfinkel and J. Stallaert (2005). "A Market Design for Grid Computing." Working Paper.

[39] Schnizler, B., D. Neumann, D. Veit and C. Weinhardt (2004a). A Multiattribute Combinatorial Exchange for Trading Grid Resources. 13th Research Symposium on Emerging Electronic Markets (RSEEM 2005), Amsterdam, forthcoming.

[40] Schnizler, B., D. Neumann, D. Veit and C. Weinhardt (2005b). Designing a Combinatorial Exchange for Computational Resources. Working Paper.

[41] Shneidman, J., C. Ng, D. C. Parkes, A. AuYoung, A. C. Snoeren, A. Vahdat and B. Chun (2005a). Why Markets Could (But Don't Currently) Solve Resource Allocation Problems in Systems. 10th USENIX Workshop on Hot Topics in Operating Systems (HotOS-X), Santa Fe, NM.

5 Grid at the Interface of Industry and Research

M. Resch

Grid environments are typically designed for a specific field of application. We see on the one hand Grids in which a certain group of scientists collaborate. A good example is the large hadron collider Grid (LHC-Grid) [17]. On the other hand there are internal Grids in industry that meet the requirements of a certain company only. The key challenge lies in combining the two communities to create a Grid that is truly interdisciplinary and ubiquitous. This paper describes an approach that has been taken at the Höchstleistungsrechenzentrum Stuttgart (HLRS) and has proven to be successful over the last ten years.

5.1 What is a Grid?

Before setting off to describe the usage of Grids in industry and research we would like to get a better understanding of what "Grid" actually means. The term „Grid" has never been clearly defined in literature although a number of papers and books have claimed to give an overview of the Grid [7][9]. Having a look at the history of the Grid one may, however, get a better understanding of the background of the topic.

In the mid 90s of the last century supercomputer centres were looking at ways to increase the performance of their systems. Parallel computing had been established as a proven technology and clusters of workstations had attracted some interest but at that time had not yet proven to be entirely successful. An extension of parallel computing and clustering was seen in metacomputing – the coupling of large systems over wide area networks [8][12]. First experiments were done in the US and in 1997 for the first time two supercomputers were connected across the Atlantic Ocean by HLRS and Pittsburgh Supercomputing Centre (PSC) [11]. These experiments proved to be heroic but not successful. Even though the bandwidth of national and international research networks was high enough the problem of latencies in the range of tens of milliseconds was never resolved.

In 1999 the term Grid was introduced by Foster and Kesselman [9] to replace metacomputing on the one hand but also to express that the community had to broaden its horizon. The new concept introduced was aiming at using not only computers but a set of resources to solve large scale problems. This could include scientific instruments as well as software and databases. However, the notion of the Grid as a utility to mainly harvest compute power was still kept [7].

Finally, in 2000 a new term was introduced that tries to combine internet technologies and traditional metacomputing by widening the scope of the Grid. E-science [13] – as it was called – was now much more about collaboration and co-operation based on resource sharing than on pure compute power.

Over the time all these approaches were called Grid computing. In this article we let the Grid be an infrastructure built from hardware and software to solve scientific and industrial simulation problems. On the one hand this does not reduce the Grid to the problem of computing. On the other hand it allows distinguishing Grid computing from the much wider field of general internet applications.

5.2 First approach to Grid Computing

HLRS started to work on the problem of distributed resources very early on with experimental settings demonstrated in the late 80s of the last century. When HLRS in 1996 was established as a federal supercomputing centre it at the same time set out to support industry with access to supercomputing systems. To do so, it formed a private-public-partnership with T-Systems and Porsche. A variety of different systems both from HLRS and from the industrial partners were made available to users from research and industry.

A number of problems immediately showed up:

- The heterogeneity of the systems made it difficult for users to migrate between individual platforms. Although a variety of systems was available providing a pool of resources, users were hesitant to migrate. The key issues were portability of codes but even as simple problems as system access were difficult to handle.

- Security was a problem. The level of security provided for scientific users was not high enough for industrial users. Systems had to be hidden behind fire walls. At the same time the typical openness of scientific settings had to be preserved.

- With an increased number of users from a variety of institutions and companies, data management became a growing problem. Considering security and confidentiality data management in a distributed setting had to be developed.

- Visualization of data became increasingly difficult. Not only did the growing size of main memory lead to larger sets of data. The distribution of users and the lack of bandwidth to transfer all data required a change in visualization concepts.

All these issues were tackled by HLRS over the last 10 years. A common concept was finally designed in 2000 when the centre had to define the requirements for its next generation supercomputer system.

5.3 The HLRS Teraflop Workbench Concept

The Teraflop Workbench Concept is designed to allow simulation as part of a process chain in distributed computing and data environments [4]. At the core of the concept is the idea that simulation is integrated in an either scientific or industrial workflow.

5.4 Steps of an Integrated Simulation Workflow

Simulation is a process that goes far beyond the simple execution of a job on a given computer. Typically it is itself part of a larger process in the creation of a product or a system. We therefore first of all have to consider that simulation is itself part of a chain of processes.

5.4.1 Process Chain Integration

Being part of a chain of processes implies that some interfaces have to exist and some boundary conditions have to be met. Such integration is typically seen in industrial environments. However, it becomes more and more important in research settings.

- Boundary conditions that relate to the people involved: Different people work on a problem at different times and different places. Coordination of these people is critical to make sure that information that is relevant to the simulation process and information that is derived from it are fed back into the chain of processes. This goes beyond the technical flow of information since it requires accumulating knowledge and expertise. A management problem by nature it can be supported by information technology but not entirely resolved.

- Boundary conditions that relate to the processes involved: As part of a process chain the simulation has to be integrated into the workflow. Since simulation still is a new and not very mature technology it has to adapt to established processes and their interaction.

- Boundary conditions that relate to the data involved: As part of the process chain simulation relies on data received from other processes involved. This typically requires transformation of data from a user centric representation (like in CAD) to a computer centric representation (as an input for a supercomputer) and back.

- Boundary conditions that relate to the hardware involved: Computers for simulation are part of a network that typically was not designed for simulation but for communication in a commercial setting. Integrating simulation in the process chain includes considering how supercomputers can be made part of a seamless hardware environment.

5.4.2 Simulation Steps

The actual simulation process itself can again be split into a number of individual steps. These are not always clearly distinct from each other. Furthermore the traditional triad of pre-processing, processing and post-processing has to give way to more interactive approaches. The main steps through which one has to go in simulation are, however, still the same:

5.4.2.1 Code Preparation

In preparing code for simulation the main issue – besides portability – is optimization for the chosen platform. This includes both the improvement of sustained performance on a single processor and the optimization of communication patterns for a given network and its topology. Optimization can only be done on the production system – or a similar smaller system.

5.4.2.2 Input Preparation

Input data and input files have to be generated. These have to be prepared and well documented in order to be able to understand and interpret the results. With growing available main memory of supercomputers input data sets grow in size. It becomes increasingly difficult to prepare them on smaller systems.

5.4.2.3 Computing

The goal of computing is not to achieve high performance but to get the required answer in acceptable time. The sustained performance is only relevant in as much as it guarantees the required low turn-around time. In addition to the turn-around time for a single user the overall optimum usage of the system is a goal that is important for the operator of supercomputers. Such optimum usage makes itself shown in the prize for the resource and has become an economic issue.

5.4.2.4 Simulation Control

Control over the running simulation can either be done directly by steering the simulation or by analyzing intermediate results that are accessible already during the simulation. Control is important to reduce the total time to achieve the desired result and to make more efficient use of expensive compute resources. Furthermore, as we will see, increasing sizes of data make it necessary to change from post-processing to an interactive understanding of the results of the simulation process.

5.4.2.5 Analysis of Results

With the growing size of main memories more fields of simulation move from simple models to more complex ones and from two-dimensional problems to three-dimensional ones. This increase in complexity and number of dimensions does not only result in larger output files. It also makes it more and more difficult for any user to understand the complex interactions that are hidden in the results. Standard visualization therefore starts to reach its limits in the same way as simple plotting of curves did 15 years ago. More complex techniques to get insight are required.

5.4.2.6 Archiving

With increasing compute speed redoing a simulation may be cheaper than archiving data. For this only input files have to be stored and a balance of computing costs and archiving costs has to be calculated. However, for a variety of fields of applications archives have a growing importance both for scientific and legal reasons.

5.4.2.7 Summary of Process Chain Integration

A traditional simulation workflow goes through all these steps sequentially or in an iteration loop. The flow of data that accompanies this workflow is growing. Therefore, it becomes more and more important to bring the human into the loop especially when large systems are used for a long time. Interactive control and/or steering of the simulation can help to avoid costs both in terms of money and time. Driven by new models for simulation in the process chain as well as by costs a number of requirements come up that have to be considered in supercomputer procurement.

5.5 Requirements

From the ideas above a number of requirements can be deducted that have to be considered. The key requirement, however, is that the idea of a stand-alone supercomputer has to be replaced by that of a workbench for simulation. Such a workbench has to integrate the necessary hardware and software and has itself to be integrated into the technical environment of a centre or an industrial organization.

5.5.1 Data

Data are at the centre since the overall process turns out to be a continuous transformation of data. The size of expected data files is in the range of Terabytes. Contrasting this with available speed of modern networks one of the key requirements is: Do not move data! As much as possible data should be kept at one single location. Transfer is too expensive and may delay time-to-answer to such an extent that usage of a local slower system is preferable to usage of a remote supercomputer.

Keeping data in one place requires a suitable file system. High speed access to it has to be possible during all steps of the simulation process. Furthermore it has to keep up with the required I/O speed in order to support the supercomputer facility. In addition the file system has to be integrated into a hierarchical storage management system. Data have to be able to move in and out between an archive and the file system. Transfer speed must not be hampered by any administrative overhead incurred from integrating file systems and hierarchical storage management systems.

In addition to these requirements the file system has to support heterogeneity. To avoid copying of data back and forth the central supercomputingand supportive systems (code development, pre-processing, and visualization) have to have equal and simultaneous high-speed access to the file system.

5.5.2 Networks

Networks are the lifelines of computing – both inside a system and between systems. It is obvious that the increase in speed of communication is currently no match for the increase in size of data sets that are produced by supercomputers. We ignore the slow improvement of latencies and bandwidths of the internal networks of parallel computers, since there is no hope of an improvement for the next years to come. There are some other

5.5 Requirements

open issues, however, that can be overcome and that help to increase the usability and performance of a simulation workbench.

First of all the integration of supercomputers and networks has to be improved. Network connectivity and state of the art network interface controllers have to be made part of any procurement. We have to avoid the common situation that a supercomputer is unable to saturate even the dreadful bandwidth of wide area networks. The gap between peak and sustained performance also exists for the network. Technologies like TCP off-load engines are required to achieve a high sustained performance.

Second the clear distinction between internal and external networks has to be overcome. Bringing visualization as well as the traditional pre-processing and post-processing closer to the actual simulation requires that resources for this process have to be more closely integrated with the supercomputer. The natural wall we have to jump to achieve this is that between different networks.

Third we have to solve the last mile problem. If we aim at keeping remote users in the loop we have to make sure that a minimum sustained bandwidth is available between the remote workstation and the supercomputer. We admit that this is a requirement that goes beyond the sphere of influence of a supercomputing centre. However, not overcoming this last-mile-problem supercomputers are reduced to giants on shaky foundations.

5.5.3 Software

In order for the new concept to materialize a software core is required that allows to integrate the various hardware components seamlessly into one single workbench. The software has to hide the complexity of the heterogeneous hardware components as well as software components. It has to be a true problem solving environment and has to go beyond the boundaries of the supercomputer system. Given that the growing number of hardware components and software components of a simulation workbench dramatically increases the complexity of the task such software can not be expected to be available off the shelf.

5 Grid at the Interface of Industry and Research

5.5.4 Summary of Requirements

None of the three sets of requirements (data, networks, software) can stand alone to solve the problem of a simulation workbench. All have to be fulfilled sufficiently to turn raw hardware into a working system.

5.6 Concept

Faced with the requirements of industrial and academic users the High Performance Computing Center Stuttgart (HLRS) has started to design a concept to set up a simulation workbench. The need to integrate industrial users adds security as another requirement. While for scientists confidentiality of data is mostly irrelevant it is a crucial problem for industry. Security does not only mean to protect data from being altered or copied. It goes as far as to hide completely the activities of an industrial customer from all other users. Not even the size and duration of a simulation job should become public because this information may hint at the status of development work. The overall concept laid out based on the requirements is described in Fig. 1.

5.6.1 File System

The core of the concept is the file system. It has to fulfil the following requirements:

- High Speed: Main memory of the supercomputer is in the range of 10 TB. Having the memory transferred to disk requires a bandwidth in the range of several 10 GB/s. Network connection to the file system has to be adapted to the architecture of the supercomputer.

- Integration into an HSM: The file system has to be seen as one level of a multi-level storage architecture. Therefore it is mandatory to have a connection to a hierarchical storage management system.

5.6.2. Integration of Heterogeneous Servers

To be able to provide a seamless workflow to the user pre-processing and visualization resources have to be integrated into the workbench. High-speed access to the core file system has to

5.6 Concept

be provided for heterogeneous platforms. An option for this is to use lustre [14]. There will be a trade off between speed and heterogeneity. However, with Linux becoming a common platform a compromise may be achieved.

Furthermore pre-processing and visualization systems have to be integrated into the network environment of the supercomputer. An optimum solution is to have these systems connect directly to the internal network of the supercomputer. This can be more easily achieved using standard components. Network technology like Infiniband [18] might be a feasible solution. This network relies on standard interfaces and is supported on a variety of different systems.

5.6.3. Integration of Visualization and Supercomputing

In order to turn the supercomputer into a workbench the user has to be able to interactively visualize results of an application or even steer the application according to her needs. Such interactive and intuitive steering of simulations requires an integrated handling environment where the engineer defines, starts, controls, and steers the execution of a simulation. The key components for such an environment are:

- Consistent data representation: In complex simulations more than one software module is used. A consistent data representation allows to automatically run a sequence of such modules.

- Automatic execution environments: Integrating simulation tools and modules into graphical environments allows defining and starting automatic sequences of simulations. In addition they provide the engineer with a higher-level view of the simulation and intermediate and final results.

- In virtual prototyping a virtual reality environment enables the engineer to have a more realistic impression of the simulation. At the same time the engineer may take a more abstract view and pure simulation can be made transparent to the benefit of increased quality of conceptual work.

In the concept presented visualization may take place either on the supercomputer, on the visualization server or at the user workstation. Depending on the size of data and quality of the network connection any of these solutions or a mixture of them can be chosen.

5.6.4 Software Integration

The software core for such a simulation workbench can not be a monolithic solution but will be created from the interaction of a number of modular tools.

A tool for the integration of visualization and simulation, which also helps to integrate the various steps of the simulation workflow, is COVISE [15]. COVISE stands for Collaborative Visualization Environment and was originally developed to allow a group of distributed scientists to work on a common problem using visualization. COVISE can provide its basic data management features as well as its ability to do the computational work on visualization locally and distribute the pictures to any site or even a number of sites at the same time.

5.7 Conclusion

An integration of Grids in industry and research poses a number of problems that are widely ignored in the scientific community. While the purpose of scientists is collaboration and an open flow of information, industry has to work in a competitive environment with information to be hidden as much as possible. The teraflop workbench at HLRS is a collection of hardware and software that is tailored to the needs of both communities. Robust solutions are chosen in order to fulfil requirements of stability. At the same time open source is a key factor in order to keep up with the progress.

5.8 References

[1] Peggy Lindner, Edgar Gabriel, Michael Resch, *GCM: A Grid configuration Manager for heterogeneous Grid environments*, International Journal of Grid and Utility Computing 2005 - Vol. 1, No.1, pp. 4-12, 2005.

[2] Holger Brunst, Edgar Gabriel, Marc Lange, Matthias S. Müller, Wolfgang E. Nagel, Michael M. Resch, "Performance Analysis of a Parallel Application in the GRID", ICCS Workshop on Grid Computing for Computational Science, St. Petersburg, Russia, June 2-4, 2003.

[3] Ulrich Ray, "The EDM Strategy of Mercedes Car Group / Development, DaimlerChrysler Electronic Datamanagement Forum 2003 – Global Engineering, Böblingen, Germany, July 16-17, 2003.

[4] Michael M. Resch, Uwe Küster, Matthias Müller, Ulrich Lang, *A Workbench for Teraflop Supercomputing*, SNA'03, Paris, France, September 22-24, 2003.

[5] Michael M. Resch, "Clusters in Grids: Power plants for CFD", In: P. Wilders, A. Ecer, J. Periaux, N. Satofuka, P. Fox (Eds.), Parallel Computational Fluid Dynamics, Practice and Theory, Elsevier, North-Holland, pp. 285-292, 2002.

[6] M. Müller, E. Gabriel and M. Resch, „A Software Development Environment for Grid-Computing, Concurrency and Computation - Practice and Experience, Vol. 14, pp. 1543-1551, 2002.

[7] I. Foster, C. Kesselmann, S. Tuecke, „The Anatomy of the Grid: Enabling Scalable Virtual Organizations, International Journal of Supercomputing Applications, 15(3), 2001.

[8] M. Resch, D. Rantzau and R. Stoy,"Metacomputing Experience in a Transatlantic Wide Area Appkoication Testbed", Future Generation Computer Systems (15) 5-6, pp. 807-816, 1999.

[9] Ian Foster and Carl Kesselman, "The Grid: Blueprint for a New Computing Infrastructure", Morgan Kaufmann, San Francisco, 1999.

[10] E. Gabriel, M. Resch, T. Beisel, and R. Keller, "Distributed Computing in a Heterogeneous Computing Environment",

In: Vassil Alexandrov, Jack Dongarra (Eds.), Recent Advances in Parallel Virtual Machine and Message Passing Interface, LNCS 1497, pp.80-188, Springer, 1998.

[11] Michael Resch, Thomas Beisel, Thomas Boenisch, Bruce Loftis, Raghu Reddy, "Performance Issues of Intercontinental Computing", Cray User Group Meeting, San Jose (CA), May 1997.

[12] T. DeFanti, I. Foster, M. E. Papka, R. Stevens and T. Kuhfuss, „Overview of the I-WAY: Wide Area Visual Supercomputing", International Journal of Supercomputing Applications, 10, pp. 123-131, 1996

[13] http://www.rcuk.ac.uk/escience/ UK E-Science Project

[14] www.lustre.org The Lustre home page

[15] U. Lang, U. Lang, J. P. Peltier, P. Christ, S. Rill, D. Rantzau, H. Nebel, A. Wierse, R. Lang, S. Causse, F. Juaneda, M. Grave, P. Haas, "Perspectives of collaborative supercomputing and networking in European Aerospace research and industry", Future Generation Computer Systems 11 (1995), pp. 419-430.

[16] www.unicore.org The UNICORE home page

[17] http://lhc-new-homepage.web.cern.ch/lhc-new-homepage/ The Large Hadron Collider home page

[18] Michael M. Resch, Peggy Lindner, Thomas Beisel, Toshiyuki Imamura, Roger Menday, Philipp Wieder, Dietmar Erwin, "GRIDWELTEN: User Requirements and Environments for GRID-Computing", DFN Mitteilungen, 26-6/2003, 13-15.

[19] http://www.infinibandta.org/home Infiniband home page

6 Grid Computing for Systems Biology

W. Wiechert, M. Haunschild, M. Weitzel, K. Nöh, E. von Lieres, A. Wahl, E. Qeli and B. Freisleben

Systems Biology is a young research discipline that aims at the understanding of complex cellular regulation networks on the platform of system theoretic concepts. It is driven by the availability of high-throughput measurement data in recent years. Mathematical modeling has become a central tool of Systems Biology because it produces comprehensive formal representations of a complex system that can be simulated on a computer. Complex biological system models are usually built in an iterative process of experimental and theoretical work. To support this model guided discovery processes, a well assorted set of computational tools is required. Among them are parameter fitting, statistical parameter analysis and uncertainty propagation, optimal experimental design, Monte Carlo methods, model selection, or systems optimization. The computational and configurational effort of all these tasks is very high. Fortunately, they all rely on two elementary operations – simulation and sensitivity analysis – and thus are well suited for parallelization. Consequently, Grid computing methods are ideally suited for computational Systems Biology environments. This contribution summarizes some recent developments, shows the potential of Grid computing for Systems Biology and discusses two illustrative examples from metabolic flux analysis and stimulus response experiments.

6.1 Introduction

6.1.1 Systems Biology

After many years of research in molecular biology, the elementary components of living cells (DNA, RNA, ribosomes, proteins, metabolites etc.) are studied in great detail [3, 5]. These components interact in a large number of cellular processes like gene translation and transcription, enzyme catalysis, gene and enzyme regulation [38]. All together they constitute the phenomenon of life (s. Fig. 1).

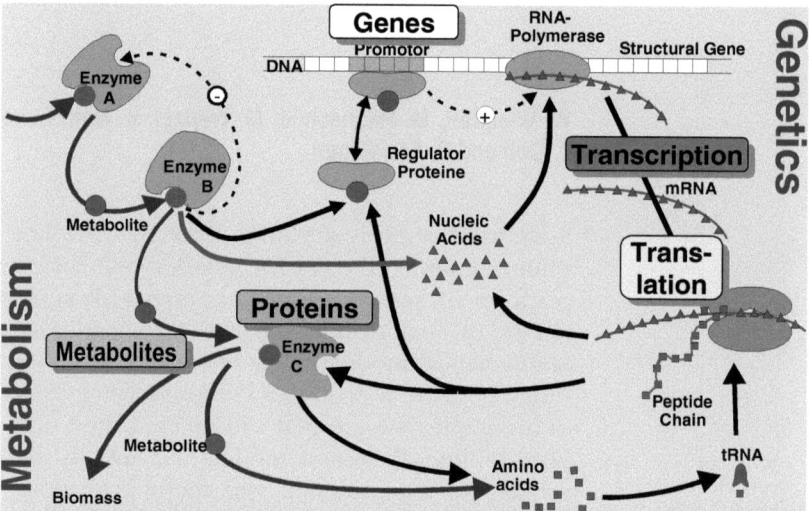

Fig. 1: Schematic representation of the major processes taking place in a living cell. The whole network – even for a simple bacterial cell – contains several thousand different units and their interactions

However, the belief was misleading that from the knowledge of the elementary components and processes it will be finally possible to reconstruct the function of the whole cellular machinery. The reason is that the complex dynamics of the cellular regulation networks were underestimated and quantitative aspects have been neglected in the reductionistic approach.

In fact, the complex behavior and tremendous diversity of living organisms results from the potential interactions between thousands of intracellular functional units. In particular, quantitative – as opposed to structural – data is now needed to assess the

6.1 Introduction

strength of these interactions and their possible influence on the whole system. In many cases highly dynamic signal transduction sequences are involved that cannot be understood from a purely static viewpoint [35,49]. Finally, even in times of high throughput experimental methods a lot of knowledge is still missing that is required to reconstruct all the interactions between cells and cellular components in a quantitative way.

A wealth of experimental methods became available in recent years that produce an enormous amount of measurement data characterizing the state of an organism *in vivo*. The first well known example was the development of DNA microchips for monitoring the transcriptional state of the whole genome [10]. Since then, many other techniques in genomics, proteomics, interactomics, metabolomics, or fluxomics have emerged [57]. The availability of this 'omics' data together with classical knowledge on molecular structure, pathways, enzyme kinetics, or thermodynamics gave rise to a new interdisciplinary research field called Systems Biology. Its ultimate goal is to understand the whole cellular concert from a systems theoretic viewpoint and to decipher the precise role of every cellular component.

Being a young discipline, the precise definition and methodology of Systems Biology is still under discussion [24, 31, 70, 76] and official definitions are rather vague like in [67]: "The study of biological systems taking into account the interactions of the key elements such as DNA, RNA, proteins, and cells with respect to one another. The integration of this information may be by computer." Nevertheless, Systems Biology has rapidly established itself as one of the driving forces in modern life sciences because systemic research on cellular function is often very closely related to practical applications in medicine or biotechnology.

For example, the understanding of drug effects on a cellular organism, or the unraveling of cellular misregulations leading to diseases are urgent problems [24]. The systematic improvement of a micro organism's performance in industrial production processes gave rise to the discipline of Metabolic Engineering in the 1990s [60]. Due to its qualitative approach, Metabolic Engineering can be attributed as "Applied Systems Biology". Consequently, many Metabolic Engineering tools are now transferred to general Systems Biology [61,69].

6.1.2 Systems Thinking

In general, systems are aggregates of interacting parts. Considering a real system as a system means to identify these parts and the relevant interactions. The driving force of applied systems thinking is the ancient Archimedean cognition that "The whole is more than its parts" as it was formalized in the 1950s by Norbert Wiener using mathematical terminology. Basically, Systems Biology is nothing else than the application of classical systems thinking to biology, although with some special flavor [76].

Systems thinking is always guided by a concrete question taken from the application context. Thus, the definition of the system under investigation, its parts and subsystems and their possible interactions depend on the viewpoint. Complex real objects like living cells or parts of them can be represented in many different ways as systems. Generally, a systems approach requires a strict focusing of research to some specified aspects of reality while neglecting others. In particular, Systems Biology does not mean to take all the information we can get to answer any open question about the whole cell. Nevertheless, as a result of the growing experimental data base, the underlying systems concept usually grows in the same way.

The structural representation of a system by parts and interactions can be enhanced by a quantitative description using state variables and equations. For a quantitative understanding of systems this representation is absolutely necessary because the arising mathematical models that can be simulated on a computer [69]. The models in turn enable to test the validity of quantitative hypothesis on the cellular function by experimentation.

A formal system representation typically grows up in an evolutionary process by changing and refining an initial model. With growing complexity, models can no more be handled manually and thus software tools for model building, simulation, data evaluation, statistical analysis, prediction or optimization are required [69]. In fact, modeling and simulation have become central activities in Systems Biology and are often considered as core competences [7, 24]. In practice, this leads to a tight integration of theoretical and experimental work. In an experimentation-modeling-cycle new models are formulated and used to suggest and design new experiments, which in turn produce the measurement data that is required to improve the existing models (cf.

6.1 Introduction

Fig. 2). This ongoing process must be supported by appropriate software tools.

Although both disciplines must work hand in hand in practice, Systems Biology should not be confused with bioinformatics. The latter is concerned with "the collection, classification, storage, and analysis of biochemical and biological information using computers" [67], i.e. information technology plays a key role. On the other hand, model-based tools – the term "model" is used here in the sense of a dynamic system description based on (nonlinear, differential, stochastic etc.) equations – are quite different from typical bioinformatics tools as will be explained in the following sections.

6.1.3 Central Problems of Biological Systems Modeling

Without a doubt, mathematical modeling in Systems Biology is yet another application of a well established methodology to a new subject. Moreover, modeling in biology has been practiced for many decades [16, 23, 59] although with much less available data for model validation. Consequently, "classical" models are rather small-sized compared to the large models built up nowadays [63]. On the other hand, some characteristic features can be identified that distinguish modeling in biology from the modeling of technical systems [71]:

Data uncertainty: Typical biological measurement data are extremely noisy. A precision below 10% is excellent while larger errors are rather the rule. In many cases, only a statement on the order of magnitude of a biological quantity or a logical statement (detectable or not) can be made. Moreover, since the experimental protocols are often very complicated, systematic errors can never be excluded. This is particularly the case for data sets from different origins (literature, data bases, experiments). Finally, there are still a lot of relevant intra-cellular quantities that cannot be measured at all. For example, several enzyme activities cannot be measured due to protein instability *in vitro* [47] or transport processes are difficult to study [34]. Furthermore, most processes in the genetic apparatus, like gene transcription and translation or protein degradation are still insufficiently characterized.

Unknown mechanisms: The structure of the cellular regulatory networks is still only partially known. Although the central metabolic networks are often considered as textbook knowledge, it is not clear which pathway is present in which specific organism. For example, pathways like the Entner-Doudouroff pathway, the glyoxylate shunt or certain anaplerotic reactions are not always present [13]. Even the precise reaction mechanism of well known pathways like the pentose phosphate pathway are still under discussion [65]. The facts about enzyme regulation are only known as far as they could be investigated *in vitro*. Furthermore, it is still an open question to what extend this data can be carried over to a living system [1,40]. Much less is known about the regulatory structure of the genetic subsystem and the function of many genes.

Open boundaries: A living cell has a natural boundary to the environment given by the cell wall or membrane. The number of interactions with the environment and other cells is – at least for prokaryotic cells – relatively easy to grasp. For single cell organisms, the external conditions of cellular growth and maintenance can be very well controlled with good reproducibility. Unfortunately, the whole cell is usually not subject to investigations in Systems Biology research projects. Typically, a cellular subsystem is studied guided by some special biological questions. This can be a metabolic pathway or larger parts of metabolism, protein complexes, signaling pathways or specific genetic regulatory loops. In any case, the experiments have to be performed using whole living cells, because it is still impossible to isolate parts of the cellular machinery in a test tube under realistic physiological conditions [1, 40]. Thus, if cellular subsystems are studied, it is extremely difficult to control the external conditions of the considered cell part. It must be taken into account that many influences are unknown, cannot be controlled or even measured. The situation is even worse when cellular tissues or cell-cell interactions are studied. Under these circumstances, a mathematical model can only be a comprehensive snapshot of the current knowledge that is based on many assumptions and has to be revised when new insights or measurement techniques become available. Nevertheless, predictions made by mathematical models can be used to systematically design new informative experiments.

Despite these problems, mathematical modeling is still the only scientific method that allows to assemble and to handle the pieces of a complex dynamic machinery in a comprehensive way. There is simply no other chance for knowledge acquisition

6.1 Introduction

and guidance of experimentation which even emphasizes the necessity for close cooperation between theory and experiment. This will certainly lead to a new way of experimentation in biology guided by system theoretic concepts, new modeling methodologies and specialized tools [76]. This methodology of knowledge acquisition will be called model guided discovery in the following.

6.1.4 Model Guided Discovery

Model guided discovery generally proceeds in the following steps:

1. Experimental, literature, or data base knowledge is taken to set up an initial mathematical model representing the current structural knowledge and mechanistic hypotheses on the function of the system components.

2. The model can be used for exploratory simulation of the real system. For example, the potential cellular response to altered external conditions, changed parameters or structures can be explored.

3. Experimenting with the model brings out what is currently not known about the system and which components and processes are not well understood. This gives rise to the design of new experiments which can falsify hypotheses or yield more information on uncertain quantities.

4. After the experiment has been carried out, the model is used to interpret and evaluate the measured data, which is usually done by parameter fitting and subsequent statistical analyses of the fit.

5. Large mathematical models tend to be as difficult to understand as the complex original system. For this reason, a whole toolbox of mathematical system analysis methods can be applied to develop a deeper understanding.

6. Typically, the process of model analysis (Steps 4 and 5) leads to further model revisions after which it is continued with Step 2.

In the course of model guided discovery, the structure of the model will be altered, components will be exchanged, mechanistic equations will be revised or even different alternatives of models will be considered at the same time [19]. The knowledge gained on the biological system is rather contained in the sequence of all model increments than in the final model (if such a model ever exists). In other words, new biological insight emerges from the "center" of the model guided discovery process (cf. Fig. 2) and not from a single step or method. Consequently, it is very important for supporting software tools to document the progress of modeling.

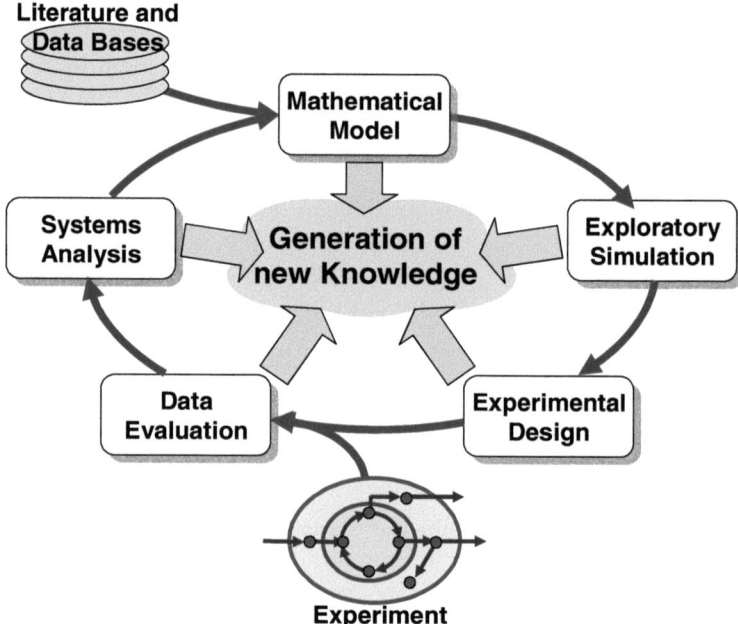

Fig. 2: *The process of model guided discovery in Systems Biology*

6.2 Grid Computing

6.2.1 Using Grid Resources for Biological Research

Most currently used models for cellular networks are based on ordinary differential equations (or similar concepts). Considering the state of the art in scientific computing [11], the numerical solution of 100-1000 equations essentially poses no problem today. This is even the case if the phenomenon of numerical stiffness

6.2 Grid Computing

occurs which is frequently the case for chemical reaction systems [11]. Additional algebraic constraints can be handled by modern DAE solvers which have been developed by extending stiff equation solvers [4]. For small- and medium-sized networks with up to one hundred reaction steps, computation times are in the order of seconds to minutes on contemporary PCs.

From the viewpoint of high performance computing this means that it makes little sense to further accelerate single simulation runs which, moreover, are hard to parallelize. Consequently, a single simulation step is identified as the most important computational primitive for all further operations.

The same statement essentially holds for sensitivity computation. If the sensitivity of a differential equation is computed with respect to one single parameter or an initial value, this can be done from the corresponding sensitivity differential equation having the same dimension as the original system [11]. Thus, sensitivity analysis with respect to one parameter is another essential building block.

Based on these elementary operations of simulation and sensitivity analysis (and a few others), many tools required for model-guided discovery, can be built up with little computational (but not necessarily implementation) overhead (s. Fig. 3). They include linearized model analysis, parameter fitting, statistical error analysis, predictions with confidence regions, Monte-Carlo methods, or model selection [69, 71]. Most of these methods rely on a repeated call to elementary operations in hierarchically nested loops. Consequently, all these methods have a natural potential for parallelization on a computer network.

Clearly, special algorithms have to be designed that can exploit the power of parallel evaluation of elementary tasks. The simplest example is sensitivity analysis with respect to all parameters which can be done in parallel on as many processors as there are parameters. If the algorithms are designed appropriately, they will also be able to work in a heterogeneous environment. However, some tools (like sensitivity analysis) are not fully scalable because there is a fixed number of steps to do.

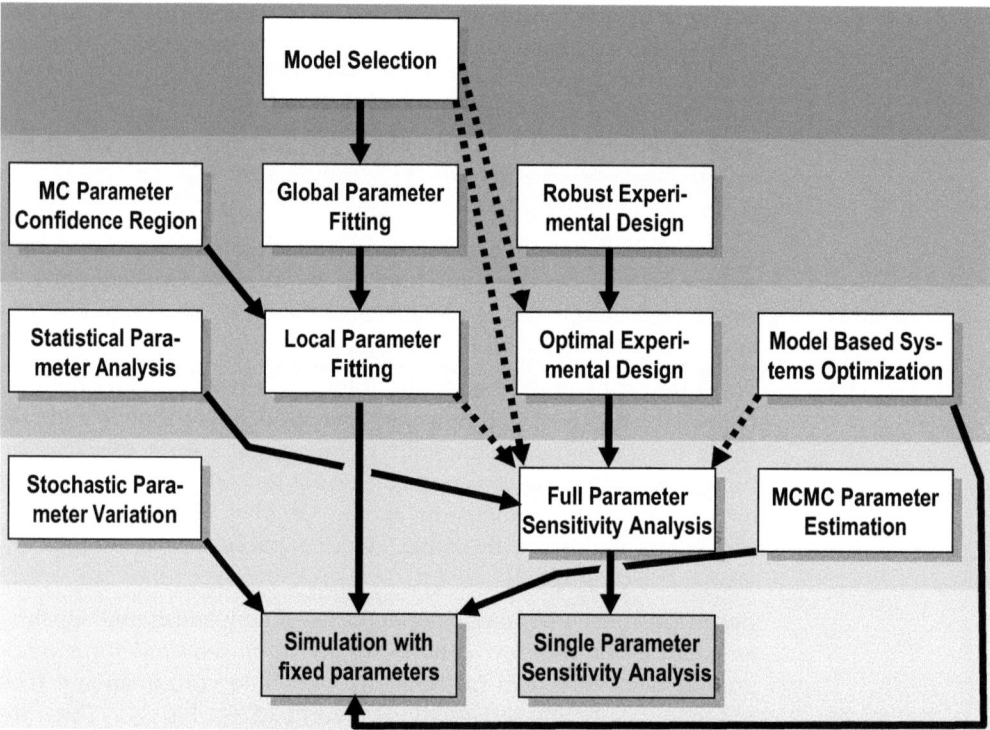

Fig. 3: Computational primitives (lowest level) and higher level algorithms for simulation, data evaluation, systems analysis, experimental design, and optimization. Each arrow represents a repeated or iterated call to a lower-level tool. The presence of the dashed arrows depends on the type of algorithm used

From the viewpoint of the software user, the computational platform should be more than just a cluster of interconnected workstations. The computing environment should additionally offer the following functionality:

- the computational primitives should be available as services in such a way that they can be initiated from any remote workstation.
- the middleware should care for resource allocation, load balancing, and fault management.
- the user should generally not be involved in the decision on which processors his or her computational tasks are actually processed.

6.2 Grid Computing

- distributed resources should be available for collaborative projects that can be used by all researchers in a project.
- well established internet facilities for user interfacing, process monitoring, data specification and exchange should be used.

All together these requirements constitute a Grid computing environment for Systems Biology.

6.2.2 Current Developments

The Systems Biology community is currently becoming aware of the Grid computing potential. The Biogrid [6] and the Life Science Grid [39] initiatives were started and a first workshop on Grid computing in the life sciences was recently organized [32]. However, most of the contributions in this workshop were rather concerned with classical bioinformatics tools than with Systems Biology applications. Bioinformatics tools are already well established on compute clusters and well accepted by the biological community. It is relatively simple to interface to a classical bioinformatics tool because only few parameters (e.g. gene or protein sequences, 3D macromolecular data, significance levels or other thresholds) have to be passed to the tool [42]. The corresponding interfaces are well documented and sometimes even standardized.

In contrast, model-based Systems Biology tools must manage heterogeneous data integration from genome to phenome. Thus, the demands on the Grid support of model-based methods are significantly higher than those for standard bioinformatics tasks. The configuration complexity of simulation tools is considerable. Model-based data evaluation requires (Fig. 4)

- model specification (with respect to structure and equations),
- model parameters and initial values,
- experimental data which can have quite different origin and format,
- measured or specified systems inputs, and
- additional parameters for the control of methods and algorithms.

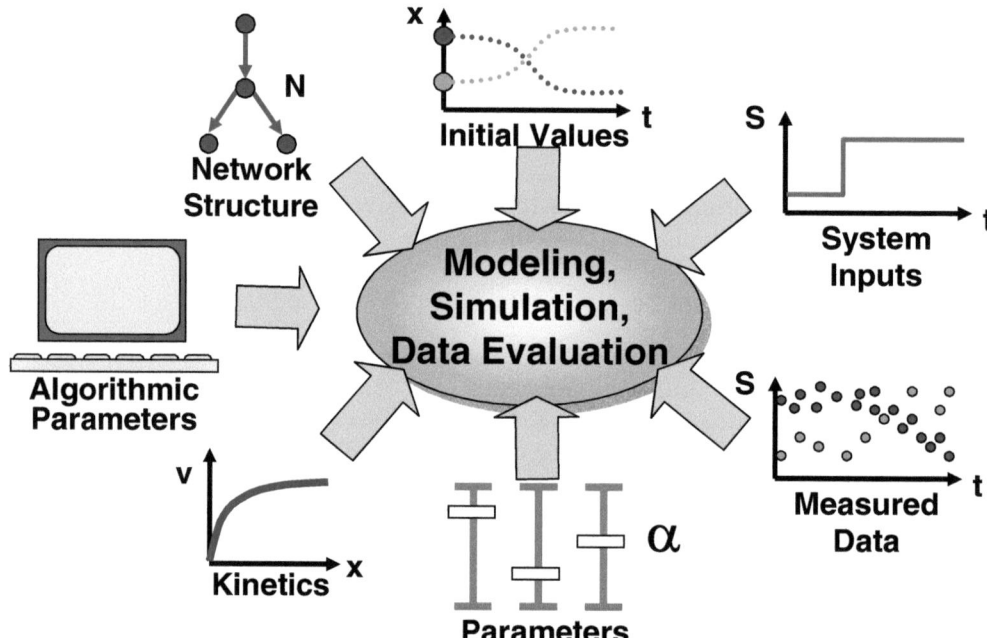

Fig. 4: Configurational data required to properly set up a biochemical network simulation for data evaluation

This configuration complexity is nothing new for a simulation specialist. In many (particularly industrial) simulation domains complexity is reduced by restricting a universal tool to certain standard application scenarios, by fixing default parameters, supplying special libraries, configuration wizards or extensive error warnings. By these measures, the majority of application scenarios can often be handled and the misuse of simulation tools by inexperienced users is prevented.

However, in the case of model-guided discovery, non-standard scenarios are rather the rule than the exception. Consequently, the full tool flexibility is required for creative work and a simulation specialist should be in the research team to exploit the full power of the tool. Only after a method has become widely accepted, it makes sense to change from universal tools to more standardized procedures [73, 78]. As an additional complication, the output of simulation tools can also be very complex and the desired information needs to be extracted first. To this end, post-processing tools and visualization aids are required and must be applied properly.

6.2 Grid Computing

The required tool flexibility is a distinguishing feature between bioinformatics and Systems Biology. Clearly, this flexibility cannot be achieved by simple command line schemes or standard data base formats but rather requires flexible mechanisms for model- and task-specification. Currently, XML-like languages like the Systems Biology markup language (SBML) [27], interfaces to scripting languages like Python and Matlab [28, 37], and code generation schemes for high performance execution [21, 29] are being developed. Both methods are well suited for Grid computing approaches. For example, the E-Cell session manager (ESM) has been implemented with Python [64]. In the same way MMT (see Sect. 6.4) uses Matlab as a scripting environment.

Currently, only a few tools for modeling and simulation in Systems Biology are Grid-enhanced and the respective applications are rather classical from the viewpoint of the scientific computing community. In fact, the methods for simulation-based optimization and parameter fitting on compute clusters have already been established in the 1990s [8, 9, 17]. Similar applications have recently been proposed for biological simulators [12, 30, 64]. Another methodology that is well established in the physical community is Monte Carlo simulation [55] that can be naturally used on a Grid for stochastic parameter variation [25, 64, 79].

Thus, only a few of the boxes in Fig. 3 are really addressed in recent literature. It must be critically assessed whether these classical applications do really justify the switch from compute clusters to a heterogeneous Grid environment. In the following, two examples from currently running research projects are used to demonstrate why Grid computing can offer many more perspectives for Systems Biology. It will be shown how the methods depicted in Fig. 3 can be practically put to work on a compute Grid and which problems must be expected thereby.

6.3 ¹³C Metabolic Flux Analysis

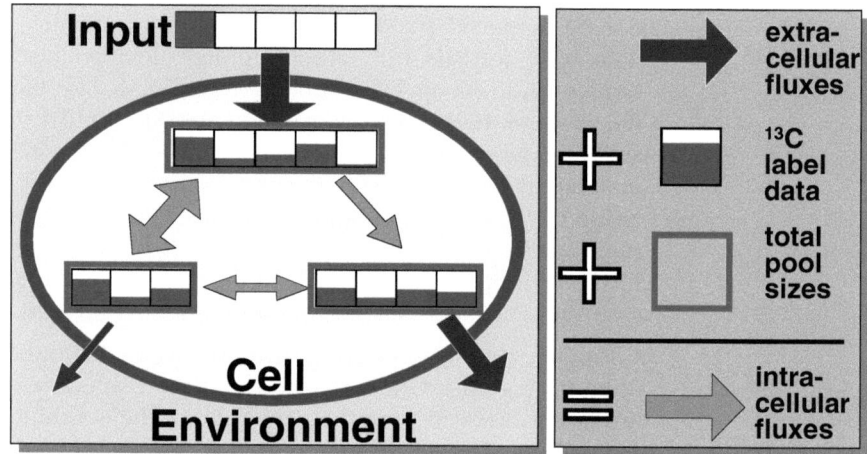

Fig. 5: *Basic functional scheme of ¹³C metabolic flux analysis. The cells are fed with a specifically ¹³C-labeled substrate (left hand side). From the measured data (right hand side) the wanted intracellular fluxes have to be computed. The total pool sizes are only necessary for isotopically instationary experiments (see text)*

6.3.1 Simulation and Sensitivity Analysis

¹³C metabolic flux analysis (MFA) is a method for determining intracellular metabolic fluxes from ¹³C isotopic labeling data and measured extra cellular fluxes (s. Fig. 5) [68]. It is a model-based data evaluation technique that derives biologically meaningful fluxome data from mass spectrometric measurement data that cannot be interpreted directly. The method is based on a complex mathematical model describing the dependency of the measured labeling data $\vec{y}$ on the unknown metabolic fluxes $\vec{v}$:

$$\text{Stoichiometry:} \quad \vec{0} = \mathbf{N}\vec{v}$$
$$\text{Label balances:} \quad \vec{0} = f(\vec{v}, \vec{x}^{inp}, \vec{x}) \quad (1)$$
$$\text{Measurements:} \quad \vec{y} = g(\vec{v}, \vec{x})$$

Here, the input data consists of the structure of the underlying metabolic network (N), assumptions on reaction irreversibilities, the carbon atom transitions in each chemical reaction step (f), the measurement configuration (g), and the known labeling state of the substrate ($\vec{x}^{inp}$). Based on this knowledge, the forward simulation step, i.e. the prediction of the measurements based on an assumed set of intercellular fluxes, constitutes the first computational primitive of MFA. Several thousands of nonlinear equations can be currently solved in the order of some seconds on a contemporary PC [72].

The evaluation of the measured data requires the solution of the inverse problem corresponding to Equation (1) with given data $\vec{y}$ and unknown fluxes $\vec{v}$. It is usually solved by a least squares parameter fit which means to put an iterative loop around the forward simulation step. The final statistical analysis of error propagations from measurements to estimated fluxes additionally requires the knowledge of the measurement sensitivities with respect to the fluxes. This is the second computational primitive of MFA. The measurement sensitivities with respect to a single flux (v_j) can be obtained by implicit differentiation of (1):

$$\vec{0} = \mathbf{N} \cdot d\vec{v}/dv_j$$
$$\vec{0} = df/d\vec{v} \cdot d\vec{v}/dv_j + df/d\vec{x} \cdot d\vec{x}/dv_j \quad (2)$$
$$d\vec{y}/dv_j = dg/d\vec{v} \cdot d\vec{v}/dv_j + dg/d\vec{x} \cdot d\vec{x}/dv_j$$

Due to the special structure of the sensitivity equations, a complete sensitivity computation with respect to all fluxes can be done with the same computational effort as one single simulation step [72].

6.3.2 High-Throughput Flux Analysis

In the classical application of ^{13}C MFA, just one experiment has to be evaluated at a time which can be easily managed on a single PC [73]. However, high-throughput MFA techniques have been recently developed that produce measurement data simultaneously for hundreds of experiments [14]. These can no longer be evaluated in interactive PC sessions. Clearly, this is a challenging task for Grid computing. It does not only include the fitting of parameters for various experiments by using Grid resources but also the diagnosis of possible errors, detection of gross data inconsistencies and separation of badly fitting models that otherwise must be identified by the human user of the flux analysis software.

Moreover, since high-throughput experiments are typically used for the screening of unknown or genetically modified micro organisms, the network topology is also not completely known (cf. Section 6.1.3). Thus, another loop around the inner parameter fitting loop should test several model alternatives in order to find the correct network topology. To summarize, this produces three nested loops (s. Fig. 3): a) model selection, b) parameter fitting, and c) forward simulation.

Clearly, by accelerating the innermost loop, the computational effort can be most effectively reduced. To this end, a new high-performance algorithm for solving Eq. (1) is currently being developed [29]. It relies on machine code generation from an analytical solution of Eq. (1) by making a sophisticated use of Kleene's classical theorem on the equivalence of finite automata and regular expressions [2]. The generated code is highly optimized, scalable on modern instruction parallel processors, and provides the fastest possible evaluation.

Code generation and distribution over a compute Grid must be managed by the middleware in this case. For this purpose, the models are formulated by the user by using a specially designed XML dialect for MFA. Based on this XML input, a code generator produces the simulation code for every single processor. First prototypes have already been tested.

6.3.3 Nonlinear Error Propagation

Another application in MFA longing for Grid computing is nonlinear error propagation. Without the consequent propagation of measurement errors, the precision of model predictions cannot be quantified [48]. Usually, the propagation of errors is investigated by linearization of the model [41]. This method is sufficient as long as there are no large errors in the measurements which are nonlinearly transduced by the model equations. In this situation, a linearized error analysis can be misleading and a more detailed analysis is necessary. Two Monte Carlo methods, which both are computationally expensive, are known to achieve this goal. Both methods profit from the new simulation algorithm (s. Sect. 6.3.2):

- The first approach works for normally distributed measurement errors. In this case, the measured data $\vec{y}$ can be superposed by another normally distributed disturbance which is randomly generated: $y'=y+\varepsilon$, $\varepsilon \in n(\mu,\Sigma)$. Then, the new data set $\vec{y}'$ is fitted. This procedure is repeated many times in a Monte-Carlo approach until the random sample of different parameter fits produced in this way gives a representation of the non-linear error transduction of the model. From this, even an exact confidence region for the parameters can be computed.

- Another approach relies on Bayesian statistics for parameter estimation. It starts with an initial guess of the parameters by supplying an a priori distribution. Then, the a posteriori distribution of the parameters is computed by a Monte-Carlo-Markov-Chain algorithm [66]. This MCMC approach does not require the solution of the inverse simulation problem but rather relies on a very large number of repetitions of the forward simulation problem. Moreover, several Markov-Chains can be run in parallel if their initial transient phase is neglected. One advantage of this method is that a priori knowledge about the system can be supplied in a natural way. It has already been demonstrated [66] that MCMC is a task that is well suited for Grid computing with metabolic systems.

6.3.4 Isotopically Instationary Experiments

So far, all mentioned labeling experiments refer to the isotopically stationary state of the system that is reached after several hours. One sample is then taken to measure the intracellular label enrichment [68]. In contrast, a novel experimental approach to ^{13}C MFA is the sampling of the intracellular labeling state in the isotopically instationary transient right after the switch to the labeled feed [74]. The advantage of this method is that the experiment only takes a time span in the order of a minute. Thus, it is possible to do an MFA for processes where metabolic stationarity (constant fluxes during the time of the experiment) is only approximately given.

The computational burden of isotopically instationary MFA is the change from the nonlinear algebraic equation system (1) to a nonlinear differential equation system. If m samples are taken this system is given by [43, 75]:

$$\vec{0} = \mathbf{N}\vec{v}$$

$$diag(\vec{X})\dot{\vec{x}} = f(\vec{v}, \vec{x}^{inp}, \vec{x}) \qquad (3)$$

$$\vec{y}(t_i) = g(\vec{v}, \vec{x}(t_i)) \quad , i = 1, 2, ..., m$$

where the metabolic pool sizes $\vec{X}$ play an important role as capacities for labeled material. Extending the classical isotopically stationary approach to the instationary case requires the solution of the corresponding flux sensitivity equation system

$$diag(\vec{X}) \, d\dot{\vec{x}}/dv_j = df/d\vec{v} \cdot d\vec{v}/dv_j + df/d\vec{x} \cdot d\vec{x}/dv_j \qquad (4)$$

which yields thousands of differential equations for realistic networks. Sensitivities must also be computed with respect to the pool sizes. Again, the solution of the sensitivity equations – possibly in conjunction with an iterative gradient based parameter fitting algorithm or an algorithm for optimal experimental design – is well suited for the distribution over a computer network.

Because the sensitivity problem for every single flux $\vec{v}_i$ already requires the solution of thousand and more equations in practice, it makes sense to further brake down the granularity of the sensitivity equations. In [44] it has been shown how this can be achieved by exploiting the special mathematical structure of Eq. (4). Roughly, it can be shown [75] that the state vector $\vec{x}$ can be decomposed into a sequence of smaller vectors

$$\vec{x} = (\vec{x}^0, \vec{x}^1, \vec{x}^2, ..., \vec{x}^k) \qquad (5)$$

in such a way that Equation (3) can be rewritten as a cascade of linear equations systems with nonlinear inhomogeneous terms:

$$diag(\vec{X}^i)\dot{\vec{x}}^i = A(\vec{v})\vec{x}^i + b^i(\vec{v}, \vec{x}^{inp}; \vec{x}^0, \vec{x}^1, ..., \vec{x}^{i-1}),$$
$$i = 0, ..., k \qquad (6)$$

A similar decomposition holds for the sensitivity equations (4). Thus, by executing each stage of the cascade on a different node of a compute cluster it is possible to break down the computation of the sensitivities to the solution of many small problems for

$$d\vec{x}^i/dv_j, \quad i = 1, ..., k, \quad j = 1, ..., l \qquad (7)$$

Here, the solution at stage *i-1* must be passed to the solution algorithm for stage *i* which is the central synchronization problem

6.3 13C Metabolic Flux Analysis

of this approach (s. Fig. 6). This problem can be solved [44] yielding a family of small problems that are well suited for compute Grids.

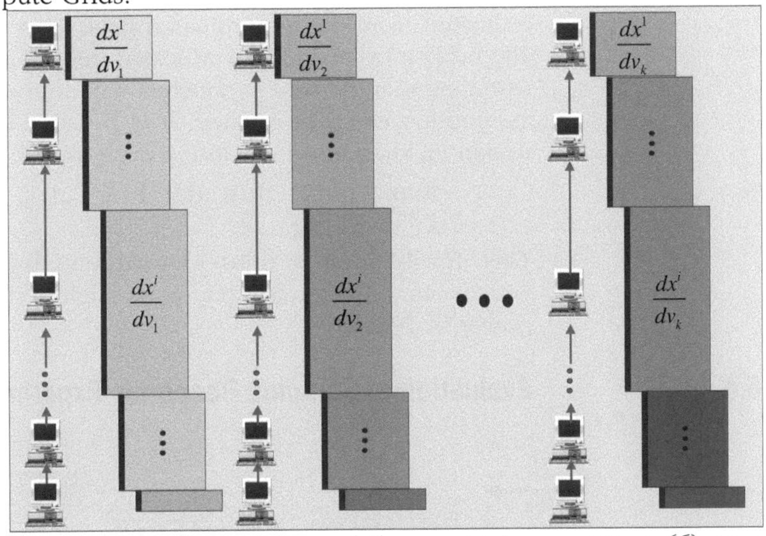

Fig. 6: Cascaded solution of the sensitivity equations (6) on a compute cluster. The distribution takes place with respect to the parameters (horizontal) and the cascade stages (vertical). Each stage needs the time lagged result of all lower stages (arrows)

6.3.5 Optimum Experimental Design

A further application of Grid computing in MFA is the computation of optimal experimental designs for isotopically stationary and instationary MFA [41, 45]. In each case, there are several experimental parameters that can be adjusted prior to the experiments. Most important are the choice of the input substrate ($\vec{x}^{inp}$) and the choice of the sampling times ($t_1, t_2, ..., t_m$) in the instationary case. The statistical determinacy of the estimated fluxes can dramatically depend on these parameters. Using a standard nonlinear D-optimal experimental design approach based on the Fisher information matrix **F** [46], this ends up with a nonlinear optimization problem

$$\max_{\vec{x}^{inp}} \max_{t_1,...,t_m} \det \mathbf{F}(\vec{x}^{inp}, t_1, ..., t_m; \vec{v}) \qquad (8)$$

The information matrix in turn is computed from the flux sensitivities (Fig. 3). Unfortunately, in the case of nonlinear experimental design, $\mathbf{F}$ also depends on the fluxes $\vec{v}$ which have to be estimated from the experimental data. This vicious circle is usually broken by making an educated guess for the parameters and assuming that the true parameters will not be too far away. If this assumption cannot be made, it is possible to compute a robust design yielding good results for every possible set of fluxes:

$$\max_{\vec{x}^{inp}} \max_{t_1,\ldots,t_m} \min_{\vec{v}} \det \mathbf{F}(\vec{x}^{inp}, t_1, \ldots, t_m; \vec{v}) \quad (9)$$

This, finally, puts a third loop around the Fisher information which can only be handled with high-performance computing power (s. Fig. 3).

6.4 Evaluation of Stimulus Response Experiments

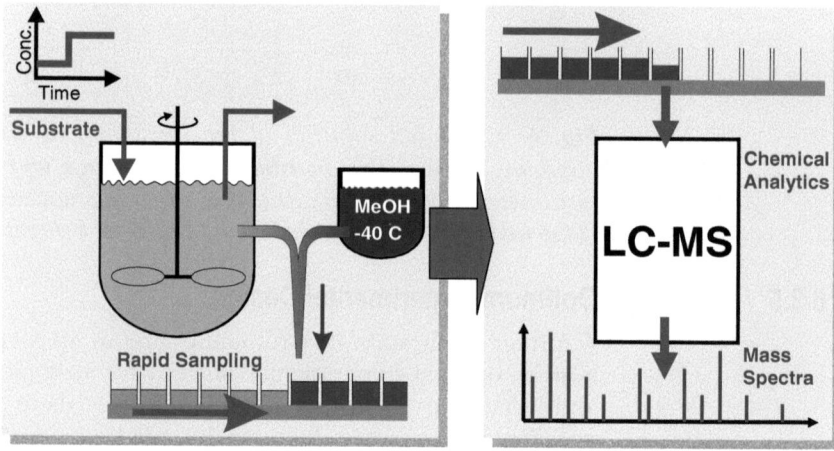

Fig. 7: Functional principle of a metabolic stimulus response experiment with rapid sampling. The cell metabolism is immediately stopped in the sample

6.4.1 Simulation and Sensitivity Analysis

Stimulus response experiments (SREs) are a promising tool to gain new information on biochemical pathways. In particular, SREs are used to verify pathway structures, to discover new regulatory mechanisms and to determine their parameters. An SRE (s. Fig. 7) starts with an intracellular stationary state of an organism. At time zero, the feed substrate concentration is abruptly raised

6.4 Evaluation of Stimulus Response Experiments

and the subsequent intracellular response is monitored by sampling, rapid stopping of cellular activity and chemical analysis of intracellular substances.

In the case of metabolic pathways in micro organisms, the response takes place at a sub-second scale and thus rapid sampling techniques are required to follow the dynamic concentration changes [33, 36, 56, 62]. One single SRE then produces several thousands concentration values from currently more than 30 intracellular pools which is a rich source of *in vivo* information on the investigated pathways.

SREs are evaluated by fitting reaction kinetic models of the investigated pathway to the measured data. The general structure of such models is given by [23, 69]

$$\text{Stoichiometry and Kinetics:} \quad \dot{\vec{X}} = \mathbf{N} \cdot \vec{v}(\vec{\alpha}, \vec{S}, \vec{X})$$
$$\text{Initial values:} \quad \vec{X}(0) = \vec{X}_0 \quad (10)$$
$$\text{Measurements:} \quad \vec{y} = g(\vec{\alpha}, \vec{S}, \vec{X})$$

The input data consists of the stoichiometry $\mathbf{N}$ of the underlying metabolic network, the metabolite concentration vector $\vec{X}$ with initial values $\vec{X}_0$, enzyme kinetic terms $\vec{v}$ with parameters $\vec{\alpha}$, known external influences $\vec{S}$, and the measurement configuration g. Again, the parameter sensitivities are obtained by implicit differentiation:

$$d\dot{\vec{X}}/d\alpha_i = \mathbf{N} \cdot \left(d\vec{v}/d\vec{\alpha} \cdot d\vec{\alpha}/d\alpha_i + d\vec{v}/d\vec{X} \cdot d\vec{X}/d\alpha_i \right)$$
$$d\vec{X}/d\alpha_i (0) = \vec{0} \quad (11)$$
$$d\vec{y}/d\alpha_i = dg/d\vec{\alpha} \cdot d\vec{\alpha}/d\alpha_i + dg/d\vec{X} \cdot d\vec{X}/d\alpha_i$$

In practice, many model variants have to be tried out, until a satisfactory fit can be achieved. This is exactly the situation of model guided discovery where models are typically not correctly describing the full measured data set.

6.4.2 Metabolic Modeling Tool

The metabolic modeling tool (MMT2) has been designed to perform all elementary computational steps of SRE evaluation (Fig. 8):

1. Model input is performed by using a specially designed XML format that was derived from SBML [18]. The graphical network editor MetVis (Metabolic Visualizer [50]) has been implemented for this purpose and is used in combination with the open source XML-editor Xerlin [77] for full specification of the model.

2. From this input and the measured data a high-performance simulation executable is generated through source code generation and compilation. This code already contains the sensitivity computation which is implemented by using automatic code differentiation based on Eq. (11) [22]. This approach is superficial to numerical differentiation schemes because no numerical approximation error occurs. Another built-in tool is network analysis based on elementary mode computation [58].

3. The simulation code can be distributed on a compute Grid using state-of-the-art Grid technologies (see below). Optimization algorithms can now orchestrate a large number of simulation tasks distributed over the network.

4. The results of simulations can finally be visualized with MetVis [50]. Moreover, Matlab is used as a flexible post-processing tool. The enormous amount of additional information produced by the sensitivity equations can be inspected by the novel graphical tool MatVis (Matrix Visualizer) based on matrix visualization, matrix reordering, multi-dimensional scaling, reorderable matrices or the Sammon mapping [51-54].

6.4 Evaluation of Stimulus Response Experiments

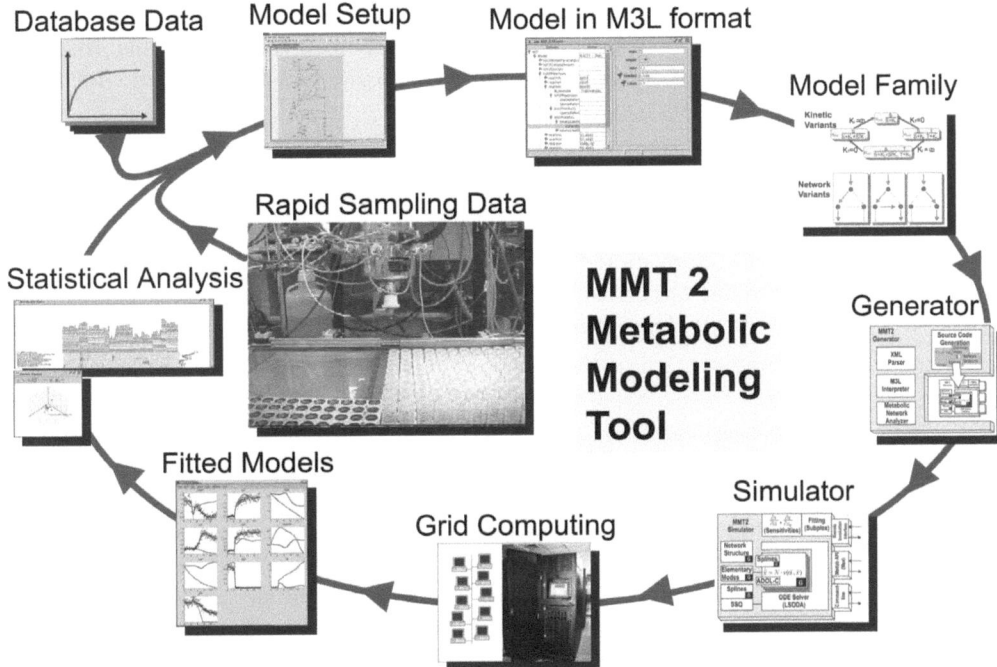

Fig. 8: Support of the model-driven discovery process for SREs by the metabolic modeling tool (MMT2)

6.4.3 External Inputs

A special mechanism in MMT accounts for the fact that the system under investigation is usually an open system having external influences. Clearly, these influences can only be incorporated if the external signals are measured. In this case, a generalized smoothing spline can be used to denoise the measured signal [18, 20]. Then the simulator works with the splined data as additional input ($\vec{S}$ in equation (10)). In particular, this concept can be used to built up a model from some few reaction steps and add more and more (measured) pools as state variables to the model. Thus, the modeler does not need to start with the full system complexity.

A major drawback of the splines is that the smoothing parameter and the chosen type of spline functions has an influence on the simulation output. In order to quantify this influence, a Monte Carlo sensitivity analysis has been performed where additional noise is added to the signal. This again has been carried out by using a distributed wrapping environment. From this analysis a realistic judgment of the data quality can be obtained (not shown here).

6.4.4 Model Selection

The most important new feature of MMT2 is the facility to specify large model families instead of just one single model [19, 20]. The underlying modeling concept replaces a single network model by a family of models

$$\dot{\vec{X}}^i = \mathbf{N}^i \, \vec{v}^i(\vec{\alpha}^i, \vec{S}, \vec{X}^i), i \in I \quad (12)$$

with an index set I [21]. The different models in a family are user-defined. They can be specified in the XML document by exchanging single reaction kinetic terms in the model. A special option is to use a Null kinetics which means to switch off a reaction step. To avoid a combinatorial explosion of the number of models in a family, an automatic feasibility check is performed based on elementary mode analysis [58] which allows to exclude biologically meaningless models [19]. Additionally, the number of models can be reduced by user-defined logical constraints.

Usually, the models in a metabolic model family are not completely different from each other but rather share many features. This induces a topology in the model space where models are linked with other models by simple alterations with respect to structure or reaction kinetics [21]. Thus, the index set is enhanced with the structure of a directed graph by introducing a relation

$$i \prec j, \quad i, j \in I. \quad (13)$$

Here, $i \prec j$ means that model j is a modification of model i in which only one alteration took place. Whenever this relation holds, two mappings

$$\pi_{j,i} : \alpha^j \to \alpha^i \quad \text{and} \quad \iota_{i,j} : \alpha^i \to \alpha^j \quad (14)$$

of the corresponding model parameter sets are specified. Hereby parameters having similar meaning in both models i and j can be identified. This additional structure makes it possible for a software tool to navigate in the space of all models [21]. If a well fitting model has been found, the parameter mappings can be

6.4 Evaluation of Stimulus Response Experiments

used as initial values for testing other models in the neighborhood.

The problem of model selection from a model family is now given by the discrete continuous optimization problem

$$\min_{i \in Feas} \min_{\vec{\alpha}^i \in Par^i} Crit(i, \vec{\alpha}^i, \vec{y}) \qquad (15)$$

where *Crit* is some statistical model selection criterion computed from the measured data and $Feas \subseteq I$ is the set of feasible models. For the i-th model, the parameter vector is $\vec{\alpha}^i$ and the space of feasible parameter vectors is Par^i.

For small model families with several hundred models, the whole model space can be exhaustively explored by using a medium-sized compute cluster. On the other hand, for large model families the computational effort is much higher, and a heuristic search strategy must be implemented. Clearly, this problem is ideally suited for parallelization on a compute Grid. Again, the simulation step is elementary.

In [21], an algorithm is presented that distributes simulation jobs over a computer network based on a multi-threaded manager worker scheme and priority queues. Model selection from a family is then solved in a sequence of nested loops:

1. The whole model space is explored by performing random trials (model sweep).

2. The neighborhood of a given model is explored by visiting similar models and making use of the similarity relation (13) (neighborhood sweep).

3. On the next level, for a fixed model, the initial values of the model parameters are randomly changed to avoid local optima (initial value sweep).

4. For each new initial value a parameter fit is triggered.

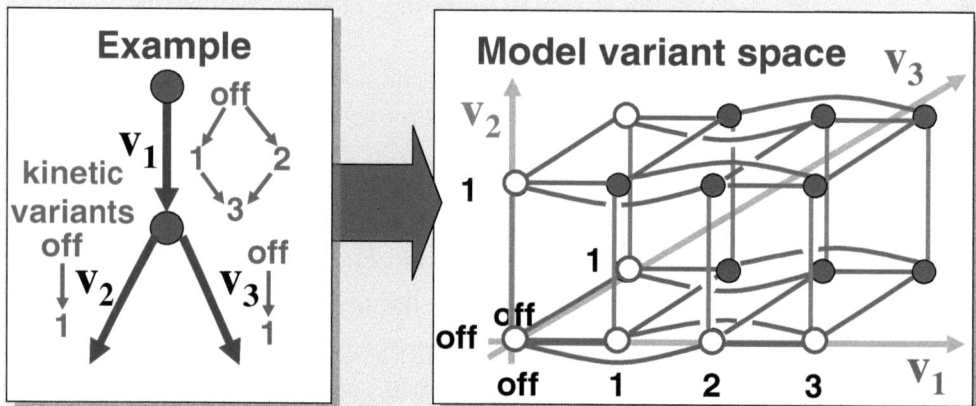

Fig. 9: *Illustration of the space of model variants induced by the model similarity relation for a simple network example (left) with kinetic variants for each reaction step. The model dependency graph (right) constructed from all kinetic combinations allows to navigate in the model space. Dark circles indicate feasible solutions while the white circles represent biologically meaningless or user excluded models [21]*

6.4.5 Technical Considerations of Grid Implementation

The technical realization of a model-based Systems Biology framework on a compute Grid is not trivial. Usually, the respective software tools exist as single-processor solutions, which are not ready for Grid computing. Moreover, they are usually composed from several components and auxiliary tools using different programming languages. In the case of MMT2 and the associated tools, C++ is mixed with raw C code and Java.

The communication between the MMT components is managed by exchanging XML or CSV files [18]. Graphical front and back ends (network setup with MetVis, XML editing with Xerlin, simulation output visualization with MetVis, sensitivity data mining with MatVis) are implemented in Java [50-54].

6.4 Evaluation of Stimulus Response Experiments

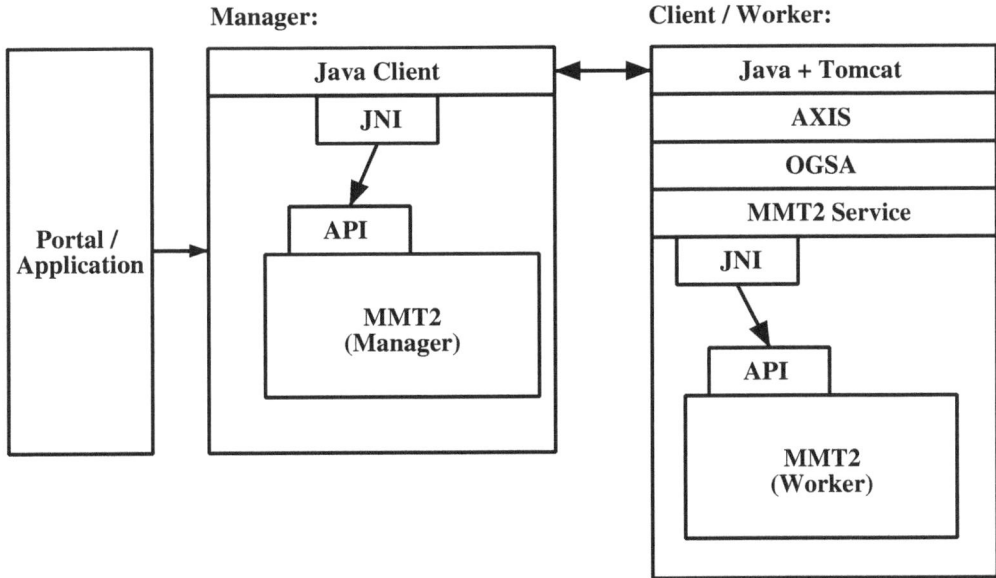

Fig. 10: *Architecture of the MMT 2 Grid computing framework*

The MMT framework was made ready for Grid computing by wrapping the MMT simulator into Grid-services using the Java-based Globus Toolkit 4.0 (GT4) as a service-oriented Grid middleware [15]. The Grid specific parts of the service are implemented as a Java interface that uses the computational methods of the MMT2 simulator. These computational methods are accessed using the Java Native Interface (JNI) since they are implemented in C. All descriptions and communication artifacts that are needed to access the MMT2 simulator as a Grid service are generated from the Java interface using a number of tools included in the GT4 distribution. The service is currently used by a basic scheduler that offers the distributed execution of the MMT2 simulator service through a web portal that is accessible using a standard web browser.

After the model has been defined by use of MetVis and Xerlin, the resulting XML description file can be submitted to the portal (s. Fig. 10, left) which then executes the MMT2 generator to build the MMT2 simulator by source code generation and compiling. The JNI needs the MMT2 simulator to be build as a DLL (dynamic link library) in case of Microsoft Windows operating systems or as a shared object in case of Linux. The generated simulator must then be distributed to the nodes, providing the MMT2 Grid service.

Computation on the Grid is realized as a manager-worker scheme, with elementary operations for simulation, sensitivity analysis and local parameter fitting. The Java implementation of the manager invokes the MMT2 simulator through the JNI for the distinct models. Each task is assigned an identifier. The computer providing the Web-Portal-Service{ XE "Web-Portal-Service" } can also be used as the manager node.

Finally, the workers providing the MMT2 Grid service are available to the manager. By sending the identifier and other simulation parameters, the computation is started, which can be queried by the manager about its progress.

The MMT 2 framework has been tested on the 256 processor (AMD Opteron) Rubens cluster of the University of Siegen and the JUMP supercomputer of the Research Center Jülich.

6.5 Conclusion

After several years of lively discussion on the implementation of Grid computing there are still new application fields which can profit from this technology. In this chapter it has been shown that Systems Biology is a challenging application domain for implementing and testing Grid technology because

1. the arising problems definitely need high performance computing power,
2. they can be parallelized on a level of granularity that does not require massively parallel computers,
3. the single tasks can be distributed over a heterogeneous Grid if the middleware and the application specific algorithms are prepared for parallel execution,
4. model based systems have a high configuration complexity, and thus a lot of nontrivial technical problems have to be solved,
5. research in Systems Biology is strictly team-oriented, and collaborative projects often spread over countries or continents.

Apart from classical bioinformatics tools, the first model-based Grid frameworks for Systems Biology are currently emerging and two examples have been extensively discussed in this contribution. It has been demonstrated how Grid technology in Systems Biology can be implemented and used in practice. Moreover, some of the major problems have been pointed out. One of them is the configuration complexity of model-based methods that must be handled appropriately.

Several typical tasks arise in model guided discovery which are parallelizable in a quite natural way. Among them are: sensitivity analysis, local parameter fitting, global parameter fitting, statistical analysis of model fits, model selection, Monte Carlo Markov Chain algorithms for nonlinear parameter estimation, stochastic algorithms for parameter variation, optimal experimental design, and model-based systems optimization. All of these problems are in some sense closely related to optimization algorithms. This emphasizes the high relevance of developing specialized Grid-enabled high-performance search algorithms for these problem classes.

6.6 References

[1] Agius, L., Sherrat, H. (Eds.), 1996. Channeling in Intermediary Metabolism (Portland Press).

[2] Aho, A.V., Hopcroft, J.E., and Ullman, J.D. (1974). The Design and Analysis of Computer Algorithms (Addison-Wesley).

[3] Alberts, B., and Johnson, A. (2002). Molecular Biology of the Cell, 4th Edition (Taylor & Francis).

[4] Ascher, U.M., and Petzold, L.R. (1998). Computer Methods for Ordinary Differential Equations and Differential-Algebraic Equations (SIAM).

[5] Berg, J.M., Tymoczko J.L., Stryer L. (2002). Biochemistry. 5th Ed. (Palgrave Macmillan).

[6] Biogrid Project. http://www.biogrid.jp/

[7] BMBF (2002). Systeme des Lebens: Systembiologie, Bundesministerium Bildung u. Forschung. http://www.bmbf.de/pub/systembiologie.pdf

[8] Boden, H., Gehne, R., and Grauer, M. (1991). Parallel nonlinear Optimization on a Multiprocessor System with distributed Memory. In Parallel Computing and Mathematical Optimization, Volume 367. (Springer), pp. 65-78.

[9] Brüggemann, F., and Grauer, M. (1991). VOpTiX - an object-oriented Environment for parallel Optimization. In Parallel Computing and Mathematical Optimization, Volume 367. (Springer), pp. 135-153.

[10] Derisi, J.L., Iyer, V.R., and Brown, P.O. (1997). Exploring the metabolic and genetic Control of Gene Expression on a genomic Scale. Science 278, 680-686.

[11] Deuflhard, P., and Bornemann, F. (2002). Scientific Computing with Ordinary Differential Equations (Springer).

[12] Dhar, P.K. (2004). The Application of Grid Technology in Systems Biology: Parameter Estimation. In: 7th Int. conference on High performance Computing and Grid in the Asia Pacific region, HPCASIA' 04. pp. 370-377.

[13] Fischer, E., and Sauer, U. (2003). A novel metabolic Cycle catalyzes Glucose Oxidation and Anaplerosis in hungry Escherichia coli. J. Biol. Chem. 278, 46446-46451.

[14] Fischer, E., and Sauer, U. (2005). Large-scale in vivo Fluxes reveal Rigidity and suboptimal Performance of B. subtilis Metabolism. Nature Genetics 37, 636-640.

[15] Foster, I., and Kesselman, C. (1997). Globus: A metacomputing Infrastructure Toolkit. Int. Journal on High Performance Computing, 115-128.

[16] Goldbeter, A., 1996. Biochemical Oscillations and Cellular Rhythms, Cambridge University Press.

[17] Grauer, M., and Pressmar, D.B. eds. (1991). Parallel Computing and Mathematical Optimization, Volume 367 (Springer).

[18] Haunschild, M. (2005). MMT2 Ref. Manual and Tutorial. http://www.simtec.mb.uni-siegen.de/software_mmt2.0.html.

[19] Haunschild, M., Freisleben, B., Takors, R., and Wiechert, W. (2005). Investigating the dynamic Behaviour of biochemical Networks using Model Families. Bioinformatics 21, 1617-1625.

[20] Haunschild, M.D., Freisleben, B., Wiechert, W., and Takors, R. (2002). Distributed Simulation of Metabolic Networks with Model Variants. In: 16th European Simulation Multiconference, ESM02, June 3-5, Darmstadt, Germany. pp. 436-440, SCS Publishing House.

6.6 References

[21] Haunschild, M.D., Wahl, S.A., Freisleben, B., and Wiechert, W. (2005). A general Framework for large Scale Model Selection. Submitted for publication.

[22] Haunschild, M.D., and Wiechert, W. (2003). Sensitivity Analysis of metabolic Network Models using Automatic Differentiation. In: 17. Symposium Simulationstechnik, ASIM 2003, Magdeburg, September 2003 (R. Hohmann, ed.). pp. 415-420, SCS-Europe.

[23] Heinrich, R., and Schuster, S. (1996). The Regulation of cellular Systems (Kluwer Academic Publishers).

[24] Henry, C.M., and Washington, C.E.N. (2003). Systems Biology. CENEAR 81, 45-55.

[25] Hill, D. (2005). Future Challenges for distributed Computing in Biology, Medicine and Biomedical Engineering. In: Modeling in Biology, Medicine and Biomedical Engineering, BioMedSim 2005: Linköping, 26-27 May 2005.

[26] Hoffmann, J.P., Ellingwood, C.D., Bonsu, O.M., and Bentil, D.E. (2004). Ecological Model Selection via evolutionary Computation and Information Theory. Journal of Genetic Programming and Evolvable Machines 5, 229–241.

[27] Hucka., M. (2003). The Systems Biology Markup Language (SBML): A Medium for Representation and Exchange of Biochemical Network Models. Bioinformatics 19, 524–531.

[28] Hunt, B.R., Lipsman, R.L., and Rosenberg, J.M. (2003). A Guide to MATLAB: for Beginners and Experienced Users (Cambridge University Press).

[29] Isermann, N., Weitzel, M., and Wiechert, W. (2004). Kleene's Theorem and the Solution of Metabolic Carbon Labeling Systems. In: German Conference on Bioinformatics, GCB 2004 (R. Giegerich and J. Stoye, eds.), vol. 53. pp. 75-84, Springer-GI.

[30] Kimura, S., Kawasaki, T., Hatakeyama, M., Naka, T., Konishi, F., and Konayaga, A. (2004). OBIYagns: A Grid-based biochemical Simulator with a Parameter Estimator. Bioinformatics 20, 1646-1648.

[31] Kitano, H. (2000). Perspectives on Systems Biology. In: New Generation Computing. pp. 199-216.

[32] Konagaya A. and K. Satou, eds. (2004). Grid Computing in Life Science, 1st Int. Workshop on Life Science Grid, LSGRID 2004, Springer.

[33] Koning, W.D., and Dam, K.V. (1992). A Method for the Determination of Changes of lycolytic Metabolites in Yeast on a sub second Time Scale using Extraction at neutral pH. Analytical Biochemistry 204, 118-123.

[34] Krämer, R. (1996). Analysis and Modeling of Substrate Uptake and Product Release by Prokaryotic and Eukaryotic Cells. Advances in Biochemical Engineering, Biotechnology 54, 31-74.

[35] Kremling, A., and Gilles, E.D. (2001). The Organization of Metabolic Reaction Networks: II. Signal-Processing in hierarchical structured functional Units. Metabolic Engineering 3, 138-150.

[36] Lange, H.C., Eman, M., Zuijlen, G.v., Visser, D., Dam, J.C.v., Frank, J., Mattos, M.J.T.d., and Heijnen, J.J. (2001). Improved rapid Sampling for in vivo Kinetics of intracellular Metabolites in Saccharomyces cerevisiae. Biotechnology and Bioengineering 75, 406-415.

[37] Langtangen, H.-P. (2004). Python Scripting for Computational Science (Springer).

[38] Lewin, B. (2004). Genes. 8th Ed. (Prentice Hall Int.).

[39] Life Science Grid. http://www.lsgrid.org/

[40] Mathews, C.K. (1993). The Cell - Bag of Enzymes or Network of Channels? Journal of Bacteriology 175, 6377-6381.

[41] Möllney, M., Wiechert, W., Kownatzki, D., and de Graaf, A.A. (1999). Bidirectional Reaction Steps in Metabolic Networks. Part IV: Optimal Design of Isotopomer Labeling Experiments. Biotechnology and Bioengineering 66, 86-103.

[42] Mount, D.M. (2004). Bioinformatics: Sequence and Genome Analysis, 2nd Edition (Cold Spring Harbour Laboratory Press).

[43] Nöh, K., and Wiechert, W. (2003). Simulation of instationary 13C labeling Experiments. In: 17. Symposium Simulationstechnik, ASIM 2003, Magdeburg, September 2003 (R. Hohmann, ed.). pp. 427-432, SCS-Europe.

6.6 References

[44] Nöh, K., and Wiechert, W. (2004). Parallel Solution Methods of cascaded ODE Systems applied to 13C-labeling Experiments. In: International Conference on Computational Science, ICCS04, Krakau, Polen, Springer.

[45] Nöh, K., and Wiechert, W. (2005). Experimental Design Principles for isotopically instationary 13C labeling Experiments. In preparation.

[46] Pazman, A. (1986). Foundations of Optimum Experimental Design (Kluwer Academic Publishing).

[47] Peters-Wendisch, P.G., Wendisch, V.F., Paul, S., Eikmanns, B.J., and Sahm, H. (1997). Pyruvate Carboxylase as an Anaplerotic Enzyme in Corynebacterium glutamicum. Microbiology 143, 1095-1103.

[48] Petersen, S., Lieres, E., Graaf, A.A.d., Sahm, H., and Wiechert, W. (2003). A Multi-scale Approach for the Predictive Modeling of Metabolic Regulation. In Metabolic Engineering in the Post Genomic Era, B.N. Kholodenko and H.V. Westerhoff, eds. (Horizon Scientific Press), chap.10.

[49] Petrov, V., Nikolova, E., and Wolkenhauer, O. (2005). A Driver Identification of the Ras/Raf/MEK/ERK Signal Transduction Pathway. Comptes Rendus d. l' Acad. Bulg. Sci. Accepted for publication.

[50] Qeli, E., Freisleben, B., Wahl, A., Degenring, D., and W. Wiechert (2003). MetVis: A Tool for Designing and Animating Metabolic Networks. In: European Simulation and Modeling Conference, ESM'2003. pp. 333-338: Naples, Italy.

[51] Qeli, E., Wiechert, W., and Freisleben, B. (2004). Visualization of Sensitivity Matrices generated during Simulations of Metabolic Network Models. In: International Conference on Applied Simulation and Modeling, ASM 2004, IASTED 2004. pp. 583-589, ACTA Press: Rhodos, Greece.

[52] Qeli, E., Wiechert, W., and Freisleben, B. (2004). Visualizing Time-Varying Matrices Using Multidimensional Scaling and Reorderable Matrices. In: Int. Conf. on Information Visualisation, IV 2004. pp. 561-567, IEEE Press: London, UK.

[53] Qeli, E., Wiechert, W., and Freisleben, B. (2005). The time-dependent reorderable Matrix Method for visualizing evolving tabular Data. In: IST/ SPIE International Conference on Visualization and Data Analysis, VDA 2005, SPIE: San Jose, USA.

[54] Qeli, E., Wiechert, W., and Freisleben, B. (2005). Visual Exploration of time-varying Matrices. In: IVO Conference on Information Visualization: London.

[55] Robert, C.P., and Casella, G. (2004). Monte Carlo Statistical Methods (Springer).

[56] Schaefer, U., Boos, W., Takors, R., and Weuster-Botz, D. (1999). Automated Sampling Device for Monitoring Intracellular Metabolite Dynamics. Anal. Biochem. 270, 88–96.

[57] Schena, M. (2003). Microarray Analysis (Wiley).

[58] Schuster, S., Fell, D.A., and Dandekar, T. (1999). A general Definition of metabolic Pathways useful for systematic Organization and Analysis of complex metabolic Networks. Nature Biotechnology 18, 326-332.

[59] Segel, L.A. (1984). Modelling dynamic Phenomena in Molecular and Cellular Biology (Cambridge University Press).

[60] Stephanopoulos, G.N., Aristidou, A.A., and Nielsen, J. (1998). Metabolic Engineering - Principles and Methodologies (Academic Press).

[61] Stephanopoulos, G., and Koffas, M. (2005). Strain Improvement by metabolic Engineering: Lysine Production as a Case Study for Systems Biology. Current Opinion in Biotechnology 16, 361-366.

[62] Theobald, U., Mailinger, W., Baltes, M., Rizzi, M., and Reu, M. (1997). In vivo Analysis of metabolic Dynamics in Saccharomyces cerevisiae: I. Experimental Observations. Biotechnology and Bioengineering 55, 305-316.

[63] Tomita, M., Hashimoto, K., Takahashi, K., Shimizu, T.S., Matsuzaki, Y., Miyoshi, F., Saito, K., Tanida, S., Yugi, K., Venter, J.C., and III, C.A.H. (1999). E-CELL: a Software Environment for whole-cell Simulation. Bioinformatics 15, 1.

6.6 References

[64] Sugimoto, M., Takahashi, K., Kitayama, T., Ito, D., and Tomita, M. (2004). Distributed Cell Biology Simulations with E-Cell System. In: Grid Computing in Life Science, 1sr Int. Workshop on Life Science Grid, LSGRID 2004, Kanazawa, Japan, May 31-June 1, 2004 (A. Konagaya and K. Satou, eds.). pp. 20-31, Springer.

[65] van Winden, W.., Verheijen, P.J.T., and Heijnen, S. (2001). Possible Pitfalls of Flux Calculations based on 13C-labeling. Metabolic Engineering 3, 151-162.

[66] von Lieres, E., and Wiechert, W. (2004). Bayes Statistics and Markov Chain Monte Carlo Simulation: An alternative Method for Parameter Identification and Error Estimation. In: 5th EUROSIM Congress on Modeling and Simulation: ESIEE Paris, Marne la Vallée, France.

[67] Webster (2002). Merriam-Webster's Medical Dictionary (Merriam-Webster).

[68] Wiechert, W. (2001). 13C Metabolic Flux Analysis. Metabolic Engineering 3, 195-206.

[69] Wiechert, W. (2002). Modeling and Simulation: Tools for Metabolic Engineering. Journal of Biotechnology 94, 37-63.

[70] Wiechert, W. (2004). Systembiologie: Eine interdisziplinäre Herausforderung. Nordrhein-Westfälische Akademie der Wissenschaften N460.

[71] Wiechert, W. (2004). Validation of Metabolic Models: Concepts, Tools, and Problems. In Metabolic Engineering in the Post Genomic Era, B.N. Kholodenko and H.V. Westerhoff, eds. (Horizon Bioscience), p. chap.11.

[72] Wiechert, W., Möllney, M., Isermann, N., Wurzel, M., de Graaf, A.A. (1999). Bidirectional Reaction Steps in Metabolic Networks. Part III: Explicit Solution and Analysis of Isotopomer Labeling Systems. Biotechnology and Bioengineering 66, 69-85.

[73] Wiechert, W., Möllney, M., Petersen, S., and de Graaf, A.A. (2001). A universal Framework for 13C Metabolic Flux Analysis. Metabolic Engineering 3, 265-283.

[74] Wiechert, W., and Nöh, K. (2005). From stationary to instationary Metabolic Flux Analysis. Adv.Biochem. Eng. Biotechnol. 92, 145-172.

[75] Wiechert, W., and Wurzel, M. (2001). Metabolic Isotopomer Labeling Systems. Part I: Global Dynamic Behaviour. Mathematical Biosciences 169, 173-205.

[76] Wolkenhauer, O. (2001). Systems Biology: the Reincarnation of Systems Theory applied in Biology? Briefings in Bioinformatics 2, 258-270.

[77] Xerlin: Opensource Extensible XML Modeling Application. http://www.xerlin.org

[78] Zamboni, N., Fischer, E., and Sauer U., FiatFlux – a Software for intracellular Flux Analysis from 13C Glucose Experiments. BMC Bioinformatics.

[79] Zi, Z., and Sun, Z. (2005). Robustness Analysis of the IFN-ã Induced JAK-STAT Signaling Pathway. Journal of Computer Science and Technology 20, 491-495.

7 Grid-basierte Simulation für die Gießerei-Industrie

J. Jakumeit

Die Gießerei-Industrie ist durch mittelständische Unternehmen geprägt, die als Zulieferer für die Großindustrie (Autohersteller, Kraftwerkbauer) arbeiten. Die Größe der Firmen geht von kleinen und mittelständischen Unternehmen (KMUs) , bis zu weltweit operierenden Firmen mit einigen 1000 Mitarbeitern. Simulation wird bisher nur wenig und vor allem in größeren Firmen eingesetzt, da eine realistische Gießsimulation sehr anspruchsvoll ist. Zur Simulation der Formfüllung und Erstarrung währen des Abgusses sollten im Idealfall Strömung, Temperaturverteilung und mechanische Deformation und Belastung gekoppelt berechnet werden. Realistische Simulationen können daher nur auf leistungsfähigen Rechnern durch geschultes Fachpersonal durchgeführt werden.

Das Interesse an solchen Simulationen steigt, da zum einen die immer engeren Vorgaben der Abnehmer an die Gussteile nur durch ein sehr genaues Verständnis des Gießvorgangs erfüllbar sind: Gusssysteme sind häufig „black boxes", die erst durch die Simulation durchsichtig und verständlich werden. Zum zweiten wächst der Druck von der Großindustrie auf die Gießereien, Simulation als Hilfsmittel in ihren Produktionsprozess zu integrieren. Ziel ist es, im Rahmen der virtuellen Fabrik den gesamten Produktionsprozess im Rechner abbilden zu können.

Die Grid-Technologie ist dabei das geeignete Mittel, um den mittelständischen Unternehmen im Gießereibereich durch kooperative Nutzung von virtuellen IT-Infrastrukturen Zugang zu realistischen Simulationen zu ermöglichen. Nur durch gemeinsame Nutzung von Ressourcen können sich KMUs bei kritischen Prozessen realistische Simulationen leisten.

Im folgenden Beitrag werden Szenarien für den Einsatz der Grid-Technologie in der Gießerei-Industrie entworfen. Dabei stehen Aspekte wie die kooperative Zusammenarbeit von Simulations- und Gussexperten trotz räumlicher Trennung, die Interaktion mit der Simulation und Sicherheitsaspekte im Vordergrund. Erste Schritte in diese Richtung werden an Hand einer aktuellen industriellen Anwendung aufgezeigt.

7.1 Simulation in der Gießerei-Industrie

Seit vielen tausend Jahren werden Metallteile durch Guss hergestellt. Dabei besteht die Kunst darin, das gesamte Gießsystem so zu konfigurieren, dass nach dem Abguss das Gussteil ohne makroskopische oder mikroskopische Hohlräume und mit der gewünschten Gefügequalität entsteht. Der Gießer hat viele Möglichkeiten, diesen Prozess zu beeinflussen. So muss das Angusssystem, durch das die heiße Schmelze in die Form fließt, richtig konfiguriert werden, Speiser können Vorräte von flüssiger Schmelze speichern, um beim Schrumpfen des Gussteiles während der Abkühlung Material nachliefern zu können. An anderen Stellen kann durch Metallteile oder Wasser die Form bewusst gekühlt werden. Auch die Zusammensetzung der Legierung hat einen entscheidenden Einfluss auf das Füll- und Erstarrungsverhalten und ist mitentscheidend für den Erfolg des Gusses. Ein großes Problem ist, das die Gussform eine „black box" ist, die eine Beobachtung des Füll- und Erstarrungsvorgangs nicht zulässt. Erst nach Öffnen der Form stellt sich heraus, ob der Guss erfolgreich war. Basierend auf den auftretenden Fehlern und seiner Erfahrung muss der Gießer bei Gussfehlern das Gusssystem verändern, bis der Guss erfolgreich ist. Dabei können die notwendigen Versuchsabgüsse teuer und zeitaufwendig sein. Einen sehr guten Überblick über die Gießtechnologie gibt J. Campbell [1].

7.1 Simulation in der Gießerei-Industrie

Dieses Szenario verlangt geradezu nach dem Einsatz von Simulation, die Licht in die „black box" der Gussform bringt und damit erklären kann, warum das gewählte Gusssystem zu Fehlern führt. Seit Anfang der 80er Jahre stehen dem Gießer immer komplexere Programmpakete zur Simulation des Formfüll- und Erstarrungsverhaltens von Gießprozessen zur Verfügung. Wurde anfänglich vor allem die zeitabhängige Verteilung der Temperatur berechnet und analysiert, so stellen moderne Programme eine gekoppelte Berechnung von Formfüllung und Temperaturverteilung oder Erstarrung und mechanischer Verformung zur Verfügung. Auch die Kornstruktur und das entstehende Gefüge können für kritische Bereiche berechnet werden [2]. Komplexe Phänomene wie die gleichzeitige Berechnung von Formfüllung und Deformation, Makroseigerung während der Erstarrung, Gaseinschluss und Oxydhautbildung sind Inhalt aktueller Forschungsprojekte und noch nicht als Standardsoftware verfügbar. Einen Überblick über den aktuellen Stand der Gießsimulation geben z.B. [3, 4, 5, 6, 7].

Ein Problem der Gießsimulation ist, dass die beim Formfüllen und Erstarren auftretenden Phänomene sehr komplex sind und eine realistische Simulation sehr rechenzeitaufwendig ist. Der Einsatz von Parallelrechnern kann hier die Rechenzeit entscheidend verringern, doch stehen entsprechende Rechner nur wenigen Gießereien zur Verfügung. Einfachere, schnelle Rechnung führt häufig zu Ergebnissen, die deutlich von den Beobachtungen abweichen und das Vertrauen in die Simulation eher schwächen. Dieser Aspekt ist besonders kritisch, da die Simulation von den Praktikern in den Gießereien immer noch mit Misstrauen beobachtet wird. Es ist schwierig, dem Praktiker die für die Simulation notwendigen Näherungen und die sich daraus ergebenden Konsequenzen für die Güte des Ergebnisses zu vermitteln. Eine gute Übereinstimmung der Simulationsergebnisse mit der Realität ist daher wichtig, um Vertrauen in die Simulation aufzubauen.

Trotz dieser Schwierigkeiten steigt das Interesse an Gießsimulationen, da erstens die Qualität der Ergebnisse kontinuierlich verbessert wurde und mittlerweile viele Phänomene im Rechner realitätsnah abgebildet werden können. Zum zweiten können die immer engeren Vorgaben der Abnehmer an die Gussteile nur durch ein sehr genaues Verständnis des Gießvorgangs erfüllt werden. Die über Generationen erworbenen Erfahrungen reichen dazu oft nicht mehr aus. Drittens wächst der Druck von der

Großindustrie auf die Gießereien, Simulation als Hilfsmitteln in ihren Produktionsprozess zu integrieren. Ziel ist es, im Rahmen der virtuellen Fabrik den gesamten Produktionsprozess im Computer abbilden zu können. Insgesamt ist daher die Nachfrage nach realitätsnahen belastbaren Simulationsergebnissen in der Gießereiindustrie groß und weiterhin steigend.

7.2 Szenarien für den Einsatz von Grid-Technologie

Die Grid-Technologie ist ein geeignetes Mittel, um den mittelständischen Unternehmen im Gießereibereich durch kooperative Nutzung von virtuellen IT-Infrastrukturen Zugang zu realistischen Simulationen zu ermöglichen. Nur durch gemeinsame Nutzung von Ressourcen können sich KMUs bei kritischen Prozessen realistische Simulationen leisten. Die Anschaffung einer eigenen Simulationsumgebung mit Fachpersonal ist häufig wirtschaftlich nicht darstellbar.

Ein realistisches Szenario für eine Grid-basierte Prozessoptimierung in der Gießereiindustrie könnte wie folgt aussehen:

An einem solchen Projekt können bis zu 5 Personen, Institutionen oder Arbeitsgruppen beteiligt sein, die gegebenenfalls an unterschiedlichen Standorten arbeiten und unterschiedliche Hardware zur Verfügung haben:

1. Der Gießingenieur (Gießereileiter), der ein Bauteil mit guter Gussqualität herstellen will. Er sollte Zugang zum Pre- und Postprozessor des Gießsimulationsprogramms haben.

2. Der Qualitätsingenieur, der sich mit Materialeigenschaften auskennt, Messtechnik hierfür besitzt und die Qualität von Gussstücken bestimmt. Er muss sich nur die Ergebnisse der Simulation anschauen können und braucht daher Zugang zum Postprozessor.

3. Der Berechnungsfachmann, der die Simulationssoftware gut kennt und die korrekte Definition dieser Simulation durch Anfangs- und Randbedingungen überprüft. Für ihn ist ein Zugriff auf Pre- und Postprozessor entscheidend.

4. Der Fachmann für rechnergestützte Optimierung, der die Optimierung von Prozessparametern durch effiziente Optimierungsalgorithmen unterstützt. Er braucht die Simulations-

7.2 Szenarien für den Einsatz von Grid-Technologie

software nicht im Detail zu kennen sondern bedient die Optimierungssoftware.

5. Die Institution, die die Ressourcen für die Rechenleistung, also z.B. ein dediziertes Cluster-System oder lokal vernetzte Workstations, auf denen die Simulations- und Optimierungssoftware läuft, zur Verfügung stellt. Dies kann natürlich auch durch den Zugang zu einem dezentralen Rechnerverbund geschehen.

Gießerei- und Qualitätsingenieur gehören dem gleichen Unternehmen an, sind aber häufig in verschiedenen Abteilungen beschäftigt die nicht am gleichen Standort angesiedelt sein müssen. Auch bei den zwei Simulationsspezialisten sind sehr unterschiedliche Kenntnisse gefordert und die Fachleute können aus verschiedenen Instituten kommen. Während der Berechnungsfachmann die Gießsimulationssoftware sehr gut kennen muss, ist dies beim Fachmann für Optimierung nicht nötig. Die Simulationssoftware ist über eine Schnittstelle an das Optimierungsprogramm gekoppelt. Mittels dieser Schnittstelle kann das Optimierungsprogramm durch Variation von Designparametern die Eingaben zur Simulation ändern und das Simulationsergebnis mit Hilfe von Optimierungskriterien bewerte. Der Optimierungsfachmann wählt die Algorithmen und Parameter des Optimierungsprogramms, um mit möglichst wenigen Simulationen eine gute Lösung zu finden. Die 5. Gruppe hat nichts direkt mit der Simulationsanwendung zu tun, sondern stellt die Rechenleistung zur Verfügung. Auf welchem Computer die Simulationen laufen und wer die notwendige Hardware zur Verfügung stellt, ist von der Problemstellung völlig unabhängig. Die Simulationen können z.B. dezentral auf einem Grid-Verbund von Workstations ablaufen.

Der gesamte Ablauf zur Simulation eines Gießprozesses kann in drei Phasen unterteilt werden:

1. Aufsetzen einer ersten Rechnung
2. Kalibrierung der Simulation
3. Optimierung des Gießprozesses

Die dazu notwendigen Arbeiten und daran beteiligten Gruppen oder Personen werden im Folgenden diskutiert.

Zu 1.: Aufsetzen der ersten Rechnung

Die Problemdefinition und das Aufsetzen erster Berechnungen für die Gießprozesssimulation müssen der Gießingenieur und Berechnungsfachmann im Dialog durchführen. Hierzu ist eine gleichzeitige interaktive Benutzung der Pre- und Postprozessoren für die Simulation über Grid-basierte Werkzeuge sinnvoll. Der Gießingenieur steuert vor allem Erfahrung über günstige Angusssysteme und Möglichkeiten für geometrische Veränderungen, der Berechnungsfachmann Wissen über die günstige Wahl von Anfangs- und Randbedingungen, Berechnungsgitter und numerische Parameter bei. Interaktiv werden die notwendigen Annahmen und Eingaben mit Hilfe Grid-basierter Kooperationswerkzeuge diskutiert.

Am Ende dieses Arbeitsabschnittes sollten sowohl der Gießingenieur als auch der Berechnungsfachmann das Gefühl haben, alle Möglichkeiten der Simulation optimal ausgenutzt zu haben, um die gewünschte Hilfe bei der aktuellen Problemstellung zu bekommen.

Zu 2.: Kalibrierung der Simulation

Wegen der Komplexität von Gießprozessen ist es selten der Fall, dass die erste Simulation den realen Prozess schon mit genügender Genauigkeit abbildet. Zumeist zeigt das erste Ergebnis, dass Wärmeübergänge zwischen Teilen des Gusssystems oder die Abstrahlcharakteristik von Teilen falsch eingeschätzt oder Annahmen im Simulationsmodell nicht realistisch sind. Die Simulation muss daher iterativ verfeinert werden, bis die Realität zufrieden stellend abgebildet wird. Für diese Kalibrierung müssen Gießingenieur und Berechnungsfachmann wiederholt Simulationsergebnisse interaktiv diskutieren und Eingaben entsprechend ändern. Die Kalibrierung der Gießsimulation kann durch den Optimierungsfachmann unterstützt werden, indem Gießingenieur und Berechnungsfachmann für kritische Parameter Wertegrenzen definieren und der Optimierungsfachmann dann den optimalen Wert durch numerische Optimierung bestimmt. Hier heißt optimal z.B., dass die berechneten Temperaturverläufe möglichst nahe an vorhandenen Messwerten liegen. Ergebnis dieses zweiten Arbeitsschrittes ist eine Prozesssimulation, die einen vorhandenen Gießprozess realitätsnah abbildet.

Zu 3.: Optimierung des Gießprozesses

Auf diesem Ergebnis kann nun die Optimierung des Gießprozesses aufsetzen. Optimierungsfachmann, Gießingenieur und Berechnungsfachmann variieren den Gießprozess, um den Prozess in Hinblick auf Gießbarkeit, Gefügequalität, benötigte Ressourcen, Prozesszeit oder andere Kriterien zu optimieren. Da auch bei paralleler Berechnung jede einzelne Simulation Stunden bis Tage dauert und für eine Optimierung viele Simulationen notwendig sind, wird die Interaktion zwischen den Anwendern nicht kontinuierlich geschehen. Vielmehr sollten die Beteiligten durch Nachrichten (z.B. Emails) darauf hingewiesen werden, dass die Optimierung des Gießprozesses ihre Aufmerksamkeit erfordert, weil z.B. neue Berechnungsergebnisse bereitstehen oder ein Optimierungslauf nicht konvergiert. Neue Berechnungsergebnisse werden vom Gießingenieur und Berechnungsfachmann interaktiv diskutiert und die Optimierung eventuell in eine neue Richtung gelenkt.

Am Ende dieses interaktiven Entwicklungsprozesses stehen ein klares Verständnis des Zusammenspiels der verschiedenen physikalischen Phänomene und ein optimierter Gießprozess.

Natürlich sind nicht bei jeder Anwendung alle 5 Arbeitsgruppen beteiligt und eine scharfe Trennung zwischen allen Gruppen ist auch nicht immer gegeben. Durch Variation ergibt sich eine Vielzahl von Szenarien. Im einfachsten Fall sind nur 2 Gruppen beteiligt, der Gießingenieur und der Berechnungsfachmann, der auch gleichzeitig die Rechnerressourcen zur Verfügung stellt. Dann dient das Grid vor allem der interaktiven Kommunikation zwischen Berechnungsfachmann und Gießingenieur. Im komplexen Fall sind alle 5 Gruppen räumlich getrennt und die Rechenleistung wird eventuell sogar von verschiedenen Institutionen zur Verfügung gestellt.

Für die Akzeptanz einer solchen Grid-basierten Entwicklung oder Analyse eines Gießprozesses sind drei Aspekte wichtig:

1. Das Grid stellt dezentral genügend Rechenleistung für realistische Simulationen zur Verfügung. Diese Rechenleistung muss bei Bedarf sofort verfügbar sein, lange Wartezeiten sind nicht akzeptabel.

2. Entscheidend für die Akzeptanz ist eine hohe Sicherheit beim Datenaustausch und der Datenverwaltung. Das Know

how des Gießers ist vor allem die Geometrie des Gießsystems. Daten mit solchen Geometrieinformationen, wie CAD-Daten und Berechnungsergebnisse als Datei oder Bild dürfen auf keinen Fall an Dritte geraten.

3. Für die interaktive Diskussion von Eingaben und Ergebnissen ist es notwendig, dass Pre- oder Postprozessor trotz räumlicher Trennung von mehreren Anwendern gemeinsam benutzt werden können. Die Programme für die Simulation und Optimierung, sowie die dazugehörigen Pre- und Postprozessoren müssen auf den verschiedenen, im Grid vereinten Rechnersystemen laufen.

7.3 Beispiel aus der Praxis

Zusammen mit einem industriellen Projektpartner und ACCESS e.V. (RWTH-Aachen) hat der Lehrstuhl für Wirtschaftsinformatik der Universität Siegen (Prof. Grauer) ein Grid-Projekt gestartet, das den Einsatz der numerischen Simulationstechnik als gestalterisches Werkzeug zur Unterstützung der Konstruktion und Fertigung von Gussteilen vorsieht. Ziel ist es, die Auswirkungen gießtechnologischer Maßnahmen zu bestimmen und darauf aufbauend praktische Versuche zu reduzieren, sowie Prozessentwicklungszeiten zu verkürzen. Bei der ausgewählten Problemstellung handelt es sich um ein Programm von Sandgussteilen aus Rotguss; Hauptprobleme sind Porositäten und Kaltläufe. Ziel ist es, ausgehend von den CAD-Daten der Gießerei, eine schnelle temperaturgekoppelte Formfüll- und Erstarrungssimulation als messbar nützliches Werkzeug einzuführen. Dabei kommen die zwei miteinander gekoppelten Pakete CASTS und COMET zum Einsatz. Neuentwicklungen dabei sind die erstarrungsgekoppelte Formfüllung im 2-Phasengebiet Schmelze/Luft unter Berücksichtigung der gasdurchlässigen Sandform sowie der kompressiblen Luft mit instationärer Temperaturverteilung. Für akzeptable Antwortzeiten bei der Simulation ist der Einsatz der Parallelrechner der Universität Siegen entscheidend.

In diesem Anwendungsfall arbeiten der Gieß- und Qualitätsingenieur am gleichen Ort und verwenden dieselbe Hardware- und Softwareinstallation. Zur Betrachtung der Simulationsergebnisse wird bei der Gießerei eine Grid-fähige Version des CASTS/COMET-Postprozessors ViewCASTS installiert. Auch die beiden Berechnungsfachleute für Gießsimulation und Optimierung sind beide bei ACCESS in Aachen. Hier wurde die Simulationssoftware entwickelt und alle notwendigen Simulations- und Auswerteprogramme stehen bei ACCESS zur Verfügung. Da ACCESS aber nicht über die ausreichende Hardware für eine parallele Gießsimulation verfügt, werden alle Rechnungen auf dem Rubens-Cluster der Universität Siegen durchgeführt. Die Daten liegen zentral auf diesem Rechner und können von ACCESS und der Gießerei aus bearbeitet werden. Für die interaktive Kommunikation wird der Postprozessor ViewCASTS von ACCESS oder der Gießerei auf den Ressourcen in Siegen gestartet und die Grafikausgabe sowohl auf dem Rechner der Gießerei als auch auf den bei ACCESS in Aachen geleitet, so dass beide Parteien die gleichen Ergebnisse im gleichen Betrachtungsmodus sehen.

Für den Datentransfer zwischen der Gießerei und der Universität Siegen ist eine Virtual Private Network-(VPN)-Leitung installiert, wodurch die Datensicherheit der eines Intranets nahekommt.

Im Moment befindet sich das Projekt am Ende der ersten Phase. Die Simulation der Formfüllung und Erstarrung des Gießens von 15 Teilen zeigt, dass ein Erstarren während des Formfüllens nicht stattfindet. Aber die Wahl des Angusssystems ist nicht ideal, da die die Gussteile nicht parallel sondern nacheinander gefüllt werden, wodurch sich unterschiedliche Erstarrungsbedingungen in den verschiedenen Teilen ergeben. Auch die Dimensionierung und Position der Speiser kann noch verbessert werden, um das Entstehen von Porositäten sicher auszuschließen.

7.4 Zusammenfassung

Wegen der vielen zusammenspielenden Phänomene sind realistische Simulationen von Gießprozessen komplex und rechenzeitintensiv. Für die zumeist kleinen mittelständischen Unternehmen in der Gießereiindustrie ist es oft nicht wirtschaftlich, die notwenige Infrastruktur und das Fachpersonal zu finanzieren. Grid-Technologie bietet die Möglichkeit, die Diskrepanz zwischen vorhandener IT-Infrastruktur und benötigter Rechenzeit zu überbrücken.

An einer Grid-basierten Gießsimulation können bis zu fünf Arbeitsgruppen mit unterschiedlichem Fachwissen und IT-Infrastruktur beteiligt sein, die an verschiedenen Orten arbeiten und interaktiv mittels des Grids einen Gießprozess analysieren und optimieren. In den drei Phasen einer simulationsunterstützten Analyse und Optimierung eines Gießprozesses bietet das Grid die notwendige Infrastuktur für ein effektives Zusammenarbeiten der verschiedenen Gruppe. Ein erstes Beispiel einer industriellen Anwendung zur Optimierung eines Sandgussproblems für Rotmetall macht das Potential deutlich.

7.5 Literaturverzeichnis

[1] J. Campbell: Castings, Butterworth-Heinemann, Oxford 1991.

[2] I.Steinbach, F.Pezolla, B.Nestler, M. Sesselberg, R. Prieler, G.J. Schmitz und J.L.L. Rezende, A phase field concept for multiphase systems, Physica D 94, 1996, S.135 ff.

[3] Proceedings of Modelling of casting, welding and advanced solidification processes – X, ed. By D. M. Stefanescu, J. A. Warren, M. R. Jolly and M. J.M. Krane, TMS, Warrendale, 2003.

[4] M. R. Jolly, S. Wen and J. Campbell, An Overview of Numerical Modelling of Casting Processes, Proceedings of Modelling & Simulation in Metallurgical Engineering and Materials Science, Beijing, 1996, S. 540 ff.

[5] M. Jolly, Casting simulation: How well do reality and virtual casting match? State of the art review, Int. J. Cast Metals Res., 14, 2002, S. 303-313.

[6] M. C. Flemmings: Solidification Modelling, Past and Present, Proceedings of Modelling of Casting, Welding and Advanced Solidification Processes VIII, ed. B. G. Thomas & Ch. Beckermann (Warrendale, PA, TMS, 1998, S. 1-13.

[7] Simulation von Gießereiprozessen, GIESSEREI kompakt 2/2004.

8 Service-Oriented Ad Hoc Grids

T. Friese, M. Smith and B. Freisleben

8.1 Introduction

The Grid computing paradigm [7, 14] is attracting a growing number of users developing larger distributed computing projects than ever before. The initial vision of the Grid encompasses the fusion of different high-performance computing centers into a common infrastructure that allows uniform access to those heterogeneous systems.

Currently, most Grid projects are in the hands of large research organizations, companies or governments such as the NASA (Information Power Grid) [36], the US Department of Energy & IBM (Science Grid) [47] and the European Union (EGEE) [12]. These institutions have dedicated staff that manage their Grid and configure it specifically for their needs. The installation of a production quality large scale Grid is far from trivial, making these administrators vital to the task.

If the Grid is to fulfil the vision of becoming the next-generation Internet (as described in [12, 13, 28,]), the complexity of installing and maintaining it must be reduced significantly. The Internet boom was made possible by making access to the Internet intuitive and transparent to the users. As a consequence, the number of users increased exponentially, which further increased the support for and the acceptance of the new medium.

The introduction of the service-oriented computing paradigm and the corresponding web service standards such as WSDL [11] and SOAP [49] in the field of Grid computing through the Open Grid Services Architecture (OGSA) [15, 16] is a major step towards reducing the complexity of Grid use, operation and maintenance. While the OGSA describes the higher-level architectural aspects of service-oriented Grid computing, the Web Service Resource Framework (WSRF) [11] is a fine-grained description of the infrastructure required to implement the OGSA model.

Service-oriented Grid computing offers the potential to provide a fine grained virtualization of the available resources to significantly increase the versatility of a Grid. It can be employed to create a broader user base as a catalyst for new Grid developments by extending the initial vision of the Grid - connecting the world's supercomputing centres - to also incorporate the much allow the Grid to offer the possibility of harnessing the unused CPU cycles (or other resources) of idle workstations, as found in practically every organization, by combining them on demand to spontaneously form an *ad hoc Grid* without a preconfigured fixed infrastructure.

To achieve this goal, a number of new challenges must be taken into account. For example, through the extension of the Grid by non-dedicated resources, the complexity of the Grid is greatly increased. Currently, a handful of administrators with specialist knowledge manage their Grid infrastructure, configure the separate nodes and preinstall all Grid services which are required. When a large number of nodes are added to the Grid on a dynamic basis, central administration is no longer feasible. The heterogeneity of the system is increased and the reliability of the nodes is decreased due to reboots or crashes caused by the regular users of those nodes. The system itself must be capable of coping with the dynamic topology changes of the underlying network and the heterogeneity of the nodes to form an ad hoc Grid autonomously. Security is also of vital importance to such an extended Grid system. Since the number of users within a system is increased, new security mechanisms are needed to ensure that malicious code cannot harm legitimate services running on the Grid.

In the following, we present the main problems involved in realizing a service-oriented ad hoc Grid to provide computing resources to every participant on demand. Our solutions to these problems are based on peer-to-peer node and service discovery, hot deployment and administration of services into a running system without disrupting other services already running there, added inter-service security by ensuring that each service runs within its own sandbox and has no direct access to the running code of other services, and a flexible trust management system.

The important parts of a prototypical implementation based on the Globus Toolkit 4.0 (GT4) will be described. The service-oriented ad-hoc Grid environment introduced in this chapter opens up a whole new range of resources to be tapped and expands the potential user base of the Grid paradigm significantly.

This chapter is organized as follows. In section 2, we introduce our notion of an ad hoc Grid. In section 3, we present the requirements that must be met by a service-oriented ad hoc Grid environment. Related work is discussed in section 4 Our prototypical implementation is described in section 5. Section 6 concludes the chapter and outlines areas for future research.

8.2 The Ad Hoc Grid

An ad hoc Grid is a spontaneous formation of cooperating heterogeneous computing nodes into a logical community without a preconfigured fixed infrastructure and with only minimal administrative requirements. The main goal of an ad hoc Grid is to provide computing resources on demand to every participant. The number of non-dedicated Grid nodes is much higher than in traditional Grid systems, demanding non-intrusive operation of the ad hoc Grid middleware.

Thus, our view of an ad hoc Grid environment goes beyond the preconfigured, dedicated Grid infrastructures existing today to encompass frequent dynamic additions of computational resources to the Grid. This includes workstations within organizations as well as scattered personal computers, similar to the basic idea of many distributed computing projects like SETI@Home [30].

While ad hoc Grid A encompasses transient nodes (e.g. non-dedicated workstations), it also includes dedicated high-performance computers. In contrast, ad hoc Grid B is made up solely of transient individual nodes. While ad hoc Grid A bears a greater resemblance to traditional Grid systems, ad hoc Grid B illustrates the shift to a personal Grid system, built without the resources of a large organization.

In the next section, we discuss the general challenges of building a service-oriented ad hoc Grid environment.

8 Service-Oriented Ad Hoc Grids

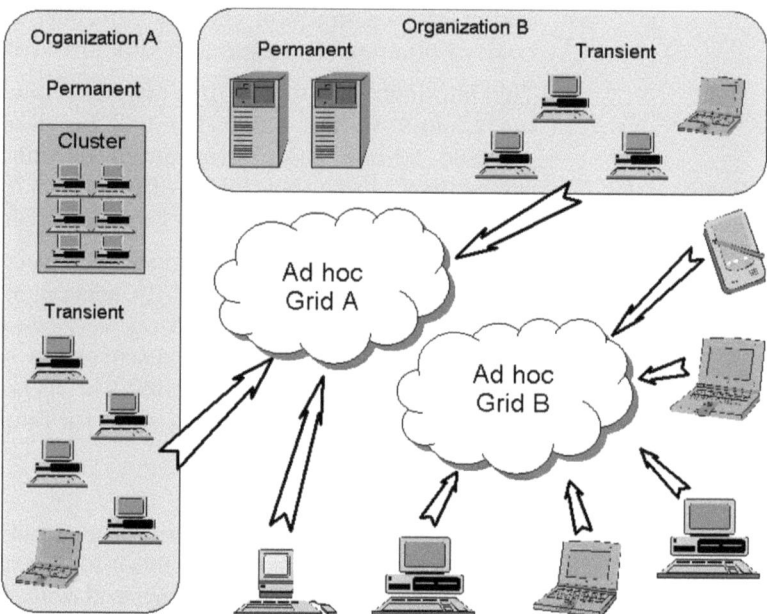

Fig. 1: Ad hoc Grid architecture overview

Fig. 1 shows how two separate ad hoc Grids are composed. The first ad hoc Grid (A) spans two organizations; the second (B) is created from scattered nodes on the Internet. Both Grid communities form a virtual organization using the existing Internet infrastructure.

8.3 Challenges

The main steps which need to be taken to allow ad hoc Grid computing in a heterogeneous environment are as follows.

8.3.1 Node Communication

Even though the web service protocols have been designed to take advantage of well known and established internet standards, some communication barriers exist on the internet that pose a problem for an ad hoc Grid environment. Private networks are hidden behind firewalls and routers performing network address translation. Nodes behind these barriers can work perfectly well as clients that consume Grid service functionality, providing services from within those confined network realms requires man-

8.3 Challenges

ual configuration effort. Approaches to support automatic traversal of such barriers are required.

8.3.2 Node / Service Discovery

In an ad hoc Grid environment, the network topology is dynamic (i.e. rebooting of workstations, movement of laptops, replacement of computers) and thus a node detection solution geared towards frequent node arrivals and departures is required. While arrival or departure of nodes should be discovered as quickly as possible, a balance needs to be found between keeping the topology information up to date and flooding the network with discovery messages.

OGSA [15, 16] and WSRF [11] define virtualization of available resources at the system-independent level of resource access to allow uniform access to a heterogeneous system. The hardware and operating system or underlying implementation of a service is hidden from the caller of a service. Support of the standard Grid service interfaces is guaranteed and sufficient to allow access to the resource. Even the instantiation of a deployed Grid service is system-independent, since it is specified to be handled by a gatekeeper (service factory).

To enable autonomous deployment of services, meta-information from the Grid node must be available to the deployment service, so it can reliably operate in a heterogeneous environment. While the underlying middleware of each node is guaranteed to support the Grid service interface, the way it implements this is not specified by OGSA or WSRF. When deploying services to newly discovered nodes, it is, however, necessary that the service implementation is supported by the Grid platform of the node. Information about the underlying hardware and operating system is particularly important when deploying legacy code, because tight integration of system resources is a common occurrence there.

Typical information needed is operating system type, available hardware resources and availability of required libraries. Furthermore, information on the reliability of the nodes can be taken into account when services are to be deployed, so nodes with long up times are given priority over nodes which frequently reboot or crash.

Furthermore, it must be made possible to discover already deployed services based on service names or service descriptions.

8.3.3 Service Deployment and Administration

Vital to the invocation of services on various nodes in a Grid system is the availability of those services. For a large scale ad hoc Grid, the time consumption for manual deployment is prohibitive, and management is difficult due to the fluctuating availability of the nodes. Even with the availability of advanced Grid programming toolkits, deployment of services has been identified as a critical issue [20]. Furthermore, the number of Grid services will steadily rise with the number of users, further increasing the management cost of the Grid environment. In a dynamically changing environment, deployment is even more critical, since there is no single deployment cycle that reaches all machines. Instead, services need to be deployed and instantiated on demand on machines as they become available.

Service deployment becomes part of an ad hoc Grid application instead of being handled by a system administrator as a precondition to the use of a service. Instead of only providing predefined services, the computational nodes of the bare Grid become a resource in themselves. This resource can be tapped by applications using the hot deployment service to leverage spare resources into their computational group. When a node becomes available that meets the requirements for the deployment of a service, the application can autonomously carry out the deployment and use the newly available machine for its application flow.

In a production environment, every operation needs to be non-intrusive, i.e. it does not interfere with the execution of other services already running on the Grid. The ability to introduce or remove a service without interruption of other operations is vital for the vision of a highly flexible ad hoc Grid.

8.3.4 Service Security

Security is a major aspect in all distributed systems, since there is always the possibility that a node introduces malicious code. In an ad hoc Grid, several new aspects must be dealt with beyond the standard security requirements existing in previous Grid systems. In traditional systems, installing a service requires security certificates allowing the operations. Services usually can only be installed by a very small number of people and trust can be assumed between all parties. In a large ad hoc Grid, on the other hand, it is possible that users unknown to each other operate on the same node. This gives rise to new security issues. One major new security threat is that a trusted node is running further unknown services. Here, inter-service security must be offered, since fair play is not guaranteed any more.

8.3.5 Service Trust

In an ad hoc Grid environment, trust is a major requirement for enabling collaboration among the interaction partners, which in our case could be nodes and/or services. Azzedin et al. [6] have classified trust into two categories: identity trust and behaviour trust. Identity trust is concerned with verifying the authenticity of an interaction partner, whereas behaviour trust deals with trustworthiness of an interaction partner.

The overall behaviour of an interaction partner consists of several elements, such as accuracy or reliability. These elements of behaviour trust should be continuously tested and verified. In this way, it is possible to collect a history of past collaborations that can be used for future decisions on further collaborations between interaction partners. This kind of experience can also be shared as recommendations to other participants.

Furthermore, the overall decision whether to trust an interaction partner or not may be affected by other non-functional aspects that cannot be generally determined for every possible situation, but should rather be under the control of the user when requesting such a decision. In addition, while the basic functionalities of two applications could be similar, differences in application behaviour could be caused by different domain specific trust requirements.

Therefore, a trust system for a service-oriented grid environment should offer flexible and easy to use components that can be configured to the specific needs of a user on a per case basis.

8.4 Related Work

There are several projects investigating a more dynamic form of Grid computing than the standard Grid solutions currently offer.

These dynamic Grids are featured under many names, e.g. ad hoc Grids, personal Grids, desktop Grids, P2P Grids, dynamic Grids. In the following, we discuss related Grid projects which focus on a dynamic Grid environment and attempt to automate the configuration and maintenance of such a Grid to enable easy adaptation and use.

In [45], a lightweight Grid environment for solving bioinformatics problems on small, "privately operated" Grids is introduced. An explicit design choice was made not to use standard Grid middleware solutions like Globus, justified by the reasons that they are too cumbersome and difficult to use in a dynamic environment and are not flexible enough to facilitate the e-science researchers' needs. Basic design criteria of the system are decentralization, provided by an underlying tuple space concept, and platform independence, provided by an implementation in Java. The following design issues are dealt with within the framework: scalability, resource allocation and scheduling, automatic distribution of application code to workers, inter-process communication and resource discovery. Furthermore, the following design issues were presented but are not yet implemented: secure resource sharing, a user billing system and quality of service mechanisms. All inter-process communication is done via Java sockets and serialized Java objects. This makes the configuration of the framework difficult to manage once organizational boundaries need to be crossed and restricts the scalability of the framework. Although this is not an issue for the scenario described in the paper, it limits the usability of the framework for other research projects. The fact that standard toolkits and protocols were avoided for the sake of usability makes it difficult to utilize components from this project or integrate new developments from other projects.

A desktop Grid computing environment for enterprise solutions is presented in [35]. Each Grid node runs a VMWare Workstation instance with Linux as the guest operating system (OS). On each guest OS, IBM's WebSphere Application Server AEs 4.0 is run to host web services which are integrated into the Grid applications. The system is targeted at enterprises which install and manage the virtual PCs and the application servers to create a small but secure Grid environment. Ease of use and manageability are not dealt with since they are not seen as critical issues.

8.4 Related Work

In [42], resource allocation algorithms are introduced which can deal with node failures in a dynamic Grid environment. In this paper, the Grid consists of mobile nodes, and the application is modelled as a directed acyclic graph. The underlying middleware is not described.

The papers [22, 21,] both discuss work in the field of medical Grid computing. The work in [22] introduces a workflow engine for medical applications with high data throughput, and [21] introduces a Grid based image analysis approach. The authors state that the integration of workstation PCs into a desktop Grid is a growing interest in the medical community. Unlike the projects mentioned above, the solutions presented in the papers are not based on lightweight custom developed Grids, but on a standard service-oriented Grid middleware. The papers state that implementing the medical applications on the standard Grid middleware is far from trivial due to the complex nature of both the application and the middleware itself.

In [3], a peer-to-peer based middleware is introduced which operates on a Bag of Tasks (BoT) in a dynamic environment where nodes can leave and join the Grid at any time. A BoT is the framework's unit of scheduling and has the following characteristics: (i) it does not need any synchronization between tasks, (ii) it has no dependencies between tasks and (iii) it can tolerate faults caused by resource unavailability with very simple strategies [4]. This places many restrictions on the application developer. The peer-to-peer system and the BoT have their own programming API and do not make use of standards.

Several criteria for ad hoc Grid computing are discussed in [2], and the need for peer-to-peer integration is explained. A figure illustrates the combination of the JXTA peer-to-peer infrastructure and the CoG [31] framework. The details of the realization of the concepts are not shown, however.

In [9], an organically inspired peer-to-peer model is introduced to facilitate the use of desktop Grids. The organic Grid is meant to bridge the gap between traditional Grids and centrally managed distributed computation projects like Seti@Home. The paper concentrates on the autonomic scheduling of simple independent tasks and does not describe the Grid architecture in which the applications run.

In [8], an approach to supporting fine-grained access control for Grid resources is proposed. The authors argue that such a fine-grained policy-based access control is necessary to enable desktop Grid computing. Giving resource owners a higher flexibility in controlling access to their resources is seen as a vital requirement for the adoption of the Grid paradigm to a higher extent into new avenues such as desktop Grids. Similarly, Grid users should get a higher flexibility in choosing the resources in which their jobs must execute in the proposal. The actual Grid architecture is not described.

A lightweight Personal Grid based on a super-node peer-to-peer network is introduced in [25]. A hierarchical scheduling system is used to distribute different types of applications in the network. The programs are described as services and are advertised by publishing XML descriptions through the peer-to-peer framework, but service-oriented standards like WSDL or SOAP are not used. Great emphasis was placed on creating a lightweight and easy to use system in contrast to the complex Grid toolkits like Globus or Unicore.

To summarize, the related work can be divided into two categories. The first deals with application specific research and the second deals with frameworks which aim to create a generic Grid environment. Developers focusing on an application often prefer a very lightweight and easy to use Grid environment [45, 42, 22, 21] and actively avoid heavyweight middleware or draw attention to the development and administrative overhead induced by the Grid middleware. The downside of this approach is that basic functionality like accounting, billing and security are only dealt with in a rudimentary fashion, and the solutions barely scale beyond the current application's scope. On the other hand, the generic Grid environment developers try to offer a feature rich environment to host many different kinds of applications at the cost of creating a steep learning curve and high administrative costs. Due to time and resource constraints, many generic middleware projects still have a specific focus of excellence [3, 9, 8, 25]. While all of the projects provide insightful research, it is very difficult to combine the research results in a single environment, due to the fact that many of these projects create their middleware from scratch. In the multi-disciplinary, multi-organizational and multi-application field of ad hoc or desktop Grids, only a cleanly designed, modular and standards based approach is able to fulfil the application developers' needs while at the same time offering the extended functionality beyond the applications' scope. The service-oriented computing paradigm offers the basis on which such a modular approach can be realized.

8.5 Implementation of a Service-Oriented Ad Hoc Grid

In this section, we present a prototypical implementation of a service-oriented ad hoc Grid, dealing with each of the challenges mentioned above.

The basis of our work is the Globus Toolkit 4.0 (GT4) deployed in Tomcat 5 [5], since it is the most widely used implementation of the Open Grid Services Infrastructure. Furthermore, since GT4 is deployed as an Axis Service in Tomcat, it offers a number of access points into the system via configuration of the Axis Service, allowing easy integration of new features.

8.5.1 P2P Infrastructure

A suitable peer-to-peer (P2P) infrastructure is beneficial for providing solutions to the problems of barrier-free node communication and for node/service discovery. Since we will refer to the P2P infrastructure used in our ad hoc Grid implementation in later sections, this section will present its properties first.

P2P systems [39, 40] are typically constructed as overlay networks to the internet infrastructure, forming a uniform virtual address and routing space over different physical networks. Their purpose is the collection of various resources in a searchable information space. One possible classification of P2P systems distinguishes between systems that use some form of flooding for search (e.g. [48]) or alternatively construct an index structure for meta-information (e.g. [46]).

Flooding describes the propagation of search requests throughout the network, assuming that every peer receiving a request acts as a relay and forwards the message to neighbouring nodes. Results matching the search request are then transmitted to the initiator of the request either through a direct connection or again by multi-hop propagation through the network. Different optimizations have been proposed, since a naïve implementation of a flooding mechanism for search requests leads to scalability problems. Those optimizations include the construction of optimized any-cast routing trees and super-peer structures. Any-cast trees are overlay routing structures that allow one-to-many broadcasts of messages from every node in the network. Super-peers are well connected nodes with a large amount of available network bandwidth that collect meta-information from many so called "edge nodes" with only limited bandwidth and connectivity. Search requests are only propagated between and answered by the super-peers, relieving the weak edge nodes from the burden of propagating search requests, so they can use their network connection for actual work.

P2P networks using flooding for search requests share two key properties. As a positive feature they provide very strong search semantics; the request is seen by every node on the network, therefore, the search can consider the entire content of the resource or associated metadata. Queries can easily use multiple wildcard patterns as long as the local search engine for resources supports them. A downside of the flooding search mechanism is the outreach of the request. In order to limit the bandwidth used for propagation of requests and results, the number of times a packet is forwarded is typically limited. This results in a so called "horizon" for every peer. There is no guarantee that a search request reaches every node in the P2P network, peers and their resources behind the horizon are invisible to a peer. This implies that answers to a search request are neither complete if results are returned nor definitive if no result is returned. There is no guarantee that ALL resources matching the request are reported and resources actually matching the request might exist behind the horizon of the requestor.

Stronger guarantees on the semantics of search operations are provided by the second class of P2P systems. These systems construct a consistent information space out of the combined resources provided by all the nodes in the network. This information space has the features of a distributed hash table (DHT) and may be used to store and retrieve key-value-pairs.

For our prototypical implementation of the ad hoc Grid we employ the Resource Management Framework (RMF) [17, 41], a P2P infrastructure developed by Siemens AG, Corporate Research, Munich, Germany. The purpose of the RMF is the creation of a flat information space for the storage and retrieval of XML documents – referred to as *resources* in the RMF. The peers

8.5 Implementation of a Service-Oriented Ad Hoc Grid

automatically form a network overlay structure that allows them to be addressed using a generalized addressing scheme. The RMF API abstracts from any details of the underlying network as well as the concrete overlay routing mechanisms used (that can be Chord, Tapestry or Napster-like) allowing a resource centric view on the information space.

The basic operations supported by the RMF are the publishing of a resource, retrieval of a resource using a given resource ID, synchronous as well as asynchronous search for resources and subscription for reception of change notification upon certain events such as the publication of a resource, its deletion from the network or changes occurring on the resource. Search and subscription operations are based on XPath queries into the contents of a resource.

The RMF handles storage of the resources on individual nodes as well as replication of resources in order to make the information space resilient against failure of individual nodes. Storage of a resource is based on resource leases. If the lease for a piece of information is not renewed by the owner of the resource within its expiration time, the resource is purged from the information space.

The basic element managed in the information space of the RMF is a *resource*. The structure of this XML element may be arbitrarily defined by an application. In order to allow the RMF resource registrar to manage this resource in the information space of the RMF, a number of child elements from the RMF registrar namespace should be present as child nodes of the root element of the resource. The most important elements of the registrar namespace are: (a) a globally unique *identifier* for the resource; the use of UUIDs is suggested in order to prevent ID collisions throughout the entire information space; (b) a user friendly *name* for the resource; this name may be used in applications for the representation of the resource regardless of the application specific content of the resource document; (c) a list of keywords that are used to identify regions of the information space where searches are conducted.

The unique ID of the resource is the only mandatory element of any resource to be published. It is possible to take virtually any XML document, include an ID as a child element of the document root and then publish the document in the RMF as a resource.

8.5.2 Node Discovery

To enable automatic service deployment in an ad hoc Grid environment, the participating nodes must be discovered first. Due to the potentially large size of future Grids, manual discovery as practiced in existing Grid environments is not an option. Since the Grid can cross the boundaries of organizations, a simple multicast or broadcast will not reach all potential participants without proper pre-configuration of the network infrastructure which is unacceptable in an ad hoc Grid environment. An automatic discovery mechanism is needed to find nodes willing to participate in the Grid. A central registry system is easy to install but does not scale well and introduces a single point of failure. For ad hoc Grids, a decentralized discovery mechanism is vital to cope with the fluctuating topology and large number of participants.

The peer-to-peer community has spent significant effort to solve the node discovery problem in large, heterogeneous and unreliable networks. Peer-to-peer and Grid computing systems have a number of similarities. Both systems aim to bring together distributed resources. In general, peer-to-peer systems are designed to fulfil a single task (e.g. file sharing), while Grids are multi-purpose and offer greater flexibility for distributed application design. The advantage of peer-to-peer systems is that they are easier to install, configure and administer. Typically, there is no central coordination needed at all. Current Grid systems are relatively small encompassing several thousands of nodes, while peer-to-peer systems can connect millions of nodes [10, 22, 32] using only the limited resources of personal computers. In [26], Iamnitchi and Foster present a more detailed comparison of the two technologies. The authors also state that peer-to-peer applications are becoming more complex, offering general distributed computing capacities. At the same time, Grid systems are growing bigger and thus the differences between the two paradigms are likely to disappear over time. Although a number of papers [26, 27, 34] discuss the benefits offered by the confluence of peer-to-peer and Grid computing, to the best of our knowledge, most of the projects suggest to integrate peer-to-peer computing

8.5 Implementation of a Service-Oriented Ad Hoc Grid

ideas only for particular aspects of Grid computing rather than integrating a fully fledged peer-to-peer solution at its core.

```
<node>
  <groupName>MarburgAdHocGroup</groupName>
  <capabilities>
    <os>
      <name>Linux</name>
      <version>Debian 3.1</version>
    </os>
    <cpu>i686</cpu>
    <memory>1024</memory>
    <peripherals>
      ...
    </peripherals>
  </capabilities>
</node>
```

Fig. 2: *A node capability record*

Our implementation of node discovery is based on the RMF described above and is aimed at providing group oriented discovery mechanisms. We assume that each of the different participants in the ad hoc Grid decides to be part of one or more collaboration groups in the system. A node registers a capability record under the group name in the RMF information space. This capability record contains static information about the node such as the operating system, processor type, total amount of installed memory and special resources available at the node (such as special sensor equipment) (s. Fig. 2).

An application requiring service deployment to other nodes can then perform a search operation in the information space, using XPath expressions to further constrain the returned node information records. As an example, an application can easily constrain the search for nodes running Linux as an operating system with the following query:

  /node/capabilities/os[name='Linux']

After gathering static information about a node, an application can directly query the node for its current state, such as the available memory and policies for the deployment of services.

8.5.3 Service Discovery

Service discovery in an ad hoc Grid as well as in a regular web service environment can happen for two purposes. Software developers in need of component services for a new application search for certain functionality, or applications already designed to consume a certain service search for available instances of the service. Grid services expose their interface descriptions as WSDL documents. This technical description - containing details about the operations exposed by a service, the data formats used for communication and protocol bindings as well as contact information such as the communication endpoints - is especially useful for applications trying to identify the right target services and determining the correct invocation methods. Standards for service registries such as the Universal Description, Discovery and Integration (UDDI) standard [29] hold provisions to also include human readable descriptions of the semantics a service offers. Other initiatives try to find ways for describing service semantics in a machine readable way for use in the vision of the semantic web [33]. Neither UDDI nor the semantic web service descriptions have seen a real breakthrough up till now. We have therefore chosen a rather pragmatic approach that leaves the possibility to implement a UDDI like search interface on top of our service description mechanism.

We have developed a service registrar on top of the underlying P2P network that is present to every node in the ad hoc Grid. This service registrar is used to map WSDL descriptions of local services into the P2P information space for later discovery and performs search operations on behalf of other applications on the ad hoc Grid. Such other applications are, for example, application schedulers in need for a particular service. The question expected to be asked most often to the service registrar is: "What endpoint addresses can be used to communicate with a service of a given port type?". We achieve this by publishing extracts of the WSDL description of a service into the information space. The port types as well as the endpoint addresses are used as keywords, allowing an efficient distributed hash table (DHT) lookup operation to be used for locating appropriate service instances.

8.5 Implementation of a Service-Oriented Ad Hoc Grid

8.5.4 Service Invocation

Grid services, like web services, have been designed to support remote invocation by sending SOAP messages to the target service. SOAP bindings to other transport protocols can be defined. The most common case is the use of HTTP as a transport protocol for SOAP messages. Most web services expose an HTTP URL as a communication endpoint in their service description that can be used to send SOAP messages in the body of an HTTP POST request to the service. In a synchronous invocation, the result is then sent back in the HTTP reply from the service to the client. HTTP has been chosen as a well known protocol that supported the tremendous success of the World Wide Web. With a very clear distinction between client and server in the communication scenario, this is certainly a good decision. First complications to this view are introduced by an asynchronous invocation pattern. In this case, the client just triggers an operation at the service that does not return a result immediately but sends the result in a second message exchange to the client. The connection to the client is initiated by the service. If the client happens to be located behind a firewall or NAT router, sending of the reply message requires the client to poll for the message, since the service cannot initiate the connection to the client.

P2P networks are designed with the basic assumption that every peer in the network must be equally accessible, even behind such barriers. Instead of relying on special implementations of NAT traversal and firewall penetration, the communication capabilities of the P2P network already used for node and service discovery can be leveraged to enable seamless communication between all nodes in the Grid system. Tunnelling of SOAP messages through the P2P network allows every node to act as a service provider with equal connectivity provided by the P2P infrastructure.

8.5.4.1 Message Handling in the Globus Toolkit

The Globus Toolkit uses the AXIS web service engine developed by the Apache Jakarta project to handle Grid service calls. AXIS is implemented as a web application that can be hosted either in a custom container distributed with the GT4 release or inside a Tomcat web application container. For the following description we assume that Tomcat is used to host the Globus web application.

The web application container handles TCP connections and decodes HTTP messages into header and body information. These request objects are then passed to the AxisServlet that acts as the entry point to the AXIS web service engine. The AXIS engine constructs a message context for every invocation that is processed by so called handler chains. Handler chains are easily configurable ordered lists of handler implementations that transform the message context. GT4 uses the WS-Addressing standard for identification of the target service to invoke; an addressing handler is configured for the request handling chain to interpret the WS-Addressing SOAP message headers and set an attribute that identifies the target service for creation of the appropriate Java classes by another handler in the request processing chain.

8.5.4.2 Tunnelling SOAP Messages through the P2P Network

There are a number of possible points for message interception in the client and message injection in the service container. Tunnel endpoints at both ends can act on the SOAP messages before they are encoded as HTTP messages, on the HTTP messages that have been constructed by the web service hosting environment or by a transparent intermediary element that can be configured as a proxy for the client. In the latter case, the HTTP request is relayed to a receiver running at the target host that acts like a regular HTTP client and issues a request to the service container. In both other cases, a component is integrated into the regular message handling infrastructure that injects the message containing the call. Combinations of the interception and injection points are also possible. For example, the HTTP message could be intercepted by a proxy implementation and then directly injected by a custom implementation of a protocol connector in the hosting environment. The different interception and injection points are depicted in Fig. 3. We have implemented the top-most connection using a custom P2P sender on the client side as well as a P2P listener that directly invokes the AXIS engine inside Tomcat.

A vital requirement of using such a message tunnelling infrastructure is the identification of the target host. Without any changes to the client or the server, the original target endpoint address for a service is the only indication for the target service.

8.5 Implementation of a Service-Oriented Ad Hoc Grid

A proxy can use the previously described service discovery component to search for a service that has been registered to use the target endpoint URL in the intercepted method. A problem of this approach is the ambiguity of the endpoint address. The Globus toolkit generates the endpoint address from the IP address of the local host. Different private networks behind NAT routers can share the same private IP address space (e.g addresses of the form 192.168.xxx.yyy). Even consideration of the service name or the service port type is likely to fail in the ad hoc Grid scenario since certain worker services are expected to be deployed on many nodes exposing the same service interface under the same IP address where the actual target nodes differ.

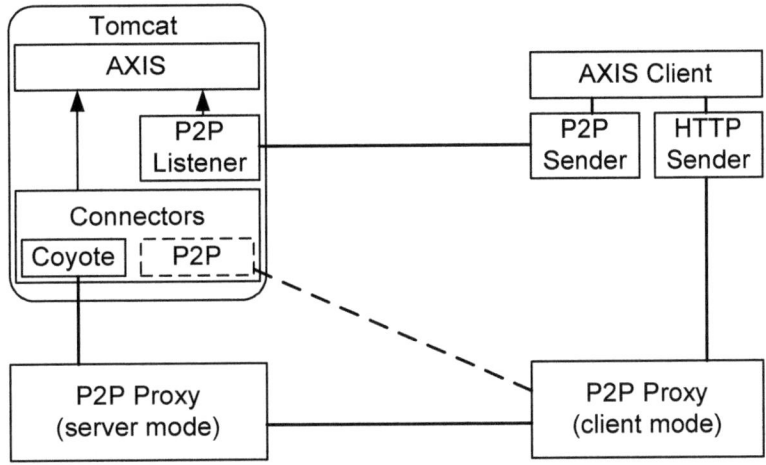

Fig. 3: Interception and injection of SOAP messages can happen at different points in the infrastructure

```
<responseFlow>
...
  <handler type="java:de.fb12....WSDLWeaveRMFHandler"/>
</responseFlow>
```

Fig. 4: Handler chain configuration

To circumvent this problem, we follow an integrated approach of early message interception and late injection, keeping the message handling chain as short as possible which also limits the amount of time needed for message encoding and decoding. A preliminary step to using this streamlined implementation is the inclusion of a custom reply handler in the global reply handler chain. Fig. 4 shows the changes that can be applied to the configuration of either the global or transport specific response handling chain.

The `invoke`-method of the class `WSDLWeaveRMFHandler` gets invoked by the engine, whenever a WSDL description of a service is generated. This method takes the WSDL document from the message context and adds a custom endpoint URL of the form `rmf://peerID/wsrf/serviceName` for the service. The client can use this target endpoint address instead of the HTTP endpoint later on.

For the actual service invocation, a listener thread is started for the Globus engine. This thread handles incoming connections from the P2P network. The message transmitted using the P2P protocol is decoded similar to the message handling performed by the HTTP connector, and handed over to the AXIS engine that handles the actual invocation of the target service instance. The reply message is then transmitted to the client using the P2P communication layer. The underlying P2P network supports TCP like connections over the multi-hop overlay network. Message injection in the platform can be configured to use access methods and content encryption and signing facilities already present in the Globus toolkit. In addition, end-to-end encryption of network connections between client and server is supported.

Support for the P2P communication infrastructure can be configured into the existing Globus infrastructure by referencing custom URL- and protocol handlers we implemented for the Java language. Registration of the custom P2P transport protocol can be achieved similar to the custom security protocol registration already performed by the standard Globus toolkit. Our ad hoc Grid toolkit allows the use of P2P communication by selecting P2P endpoint addresses without the need to ever touch any P2P specific code.

8.5.5 Service Deployment and Administration

Automatic and non-disruptive service deployment is one of the most important requirements for an ad hoc Grid. With increased adoption of the Grid paradigm, the number of developers wanting to use the Grid will increase, thus giving each developer ad-

8.5 Implementation of a Service-Oriented Ad Hoc Grid

ministrative rights for all Grid nodes is not feasible, especially in an ad hoc Grid environment where personal computers are also members of the Grid. We introduce a hot deployment service (HDS) allowing remote deployment of Grid services without requiring the developer to have an administrative account on the system. A further vital benefit of the HDS is that deployment of a service can be done non-disruptively, i.e. without requiring a restart of the WSRF platform after deployment. This is absolutely essential, since it is not acceptable that every time a new service is installed or an existing service is updated, all other services running in that environment are restarted as well - possibly losing substantial amounts of work in progress. It would be just as undesirable as to restart an entire web server every time a web page is added, but currently this is common practice in service-oriented Grid environments. If the Grid is to become the next-generation Internet, hot deployment is indispensable.

The deployment of a service currently requires a Grid service archive (GAR file) containing the needed classes, schema files and deployment descriptors that make up a service. Users of GT4 are supplied with Ant tasks that handle the distribution of the contents of this GAR file into the local standalone GT4 environment. The Ant tasks extract and copy the jar files containing the class files of the service into the local web application directory. Schema files are copied into the schema repository. The current deployment strategy of GT4 requires the restart of the entire WSRF web application, thereby killing every other Grid service currently running. Furthermore, direct access to the machine running the GT4 application is required because the Ant tasks perform all copy operations locally.

Neither the first nor the second property of the deployment mechanism is feasible for an ad hoc Grid environment with a frequently changing collection of nodes. In this environment, an application has to make sure - through dynamic service deployment - that the required service is present on every node it wishes to incorporate into its application flow.

To enable this, we modified the Axis web service engine utilized by GT4 to allow dynamic loading and unloading of Grid services. Our *hot deployment service* (HDS) provides applications with the capability to remotely *deploy, undeploy* and *redeploy* services onto a running node. The operations have the following semantics:

Deploy adds a service to the set of available services on the Grid node. The service is identified by its service name. The operation will not deploy the service if there is a service with the same name already present on the node.

Undeploy removes a service from the node, based on its service name. Running service instances already created are unaffected by the operation.

Redeploy is the chaining of undeploy and deploy. Running service instances are not changed by the redeploy operation, subsequent requests to create new instances will, however, use the newly supplied implementation of the service.

In our current implementation, access to the HDS is restricted by using the security mechanisms offered by GT4.

The basic steps the HDS needs to perform to deploy a service are:

- Register the service description with the AXIS/WSRF request handlers.
- Register the service naming description with JNDI registry.
- Make the schema files available to the WSRF environment.
- Make the service class files available to the class loader.

Currently, the need to load additional classes and dynamically replace them was not anticipated or governed by the WSRF specification or the GT4 implementation. To enable this functionality, a class loading mechanism is introduced into the realization of the HDS, as described in the following.

Grid services in GT4 are separated into three classes: The service resource class, a service home class and the service implementation itself. The service home class is used to load resources attached to a service and the service classes themselves. We provide the class HotResourceHomeImpl as our implementation of the ResourceHome interface in order to leverage our own class loading mechanism into GT4. The ResourceHome is responsible for creating the ClassLoader hierarchy which will be used to load the service classes. It distinguishes between different instances of the ClassLoaders by acquiring the service context from the AXIS engine inside the GT4 web application. It also registers all

8.5 Implementation of a Service-Oriented Ad Hoc Grid

ClassLoaders created by it at a central DisposableClassLoaderManager and the Axis ClassUtils ClassLoader cache, so they can be accessed later during undeployment. The code snippet in Fig. 5 shows the main operation of our HotResourceHomeImpl. First, the service name is extracted from the current message context. Then, the path where the jar files of the service are stored is generated based on the container configuration and an arbitrary path extension. In our case we chose basePath/WEB-INF/lib/serviceName/. Based on that, we create a JarClassLoader capable of loading all classes contained in all jar files in that directory. The JarClassLoaderManager also informs the Axis ClassUtils that it is now responsible for this service.

```
public void setResourceClass(String clazz)
    throws ClassNotFoundException {
String serviceName =
  AxisEngine.getCurrentMessageContext().getTargetService();
String basePath =
  ContainerConfig.getConfig().getInternalWebRoot();
String relPath = basePath+servicePath+libPath;
ClassLoader cl =
  JarClassLoaderManager.createLoader(serviceName, relPath)
resourceClass = cl.loadClass(clazz);
}
```

Fig. 5: *The setResourceClass operation in HotResourceHomeImpl*

This is a non-intrusive way to introduce our own class loading mechanism into GT4, since the ResourceHome implementation can be specified for each individual service. A service wishing to be hot deployable merely must use the HotResourceHomeImpl instead of the standard ResourceHomeImpl. This is the only change required to make a service hot deployable and reloadable. Hot deployable and standard services can be run side by side by using the different ResourceHome implementations. Fig. 6 shows the relationship of the ResourceHome implementations and class loaders.

The process of loading a service class is as follows. When a service is first requested, the org.globus.wsrf.jndi.BasicBeanFactory loads our HotResourceHomeImpl class in the standard Axis WebAppsClassLoader. The HotResourceHomeImpl is responsible for

8   Service-Oriented Ad Hoc Grids

creating the disposable ClassLoaders which will later load the service classes and the attached resources. When the setResourceClass method is called by the BasicBeanFactory, the CurrentMessageContext from the Axis engine is parsed to discover on behalf of which service the method is being called, thus allowing us to create one and only one ClassLoader for each service. Our JarClassLoaderManager and the modified Axis ClassUtils are informed of the service to ClassLoader mapping. Once the ResourceHome is in place, the BasicBeanFactory informs the home object which class is the main service class. As mentioned above, the HotResourceHomeImpl attaches a disposable ClassLoader to the service. Class and ClassLoader are then used by the org.globus.axis.providers.RPCProvider to instantiate the actual service object. The HotResourceHomeImpl makes sure that the class is loaded in the proper ClassLoader. Now everything is in place and the service can be accessed via the JavaProvider.

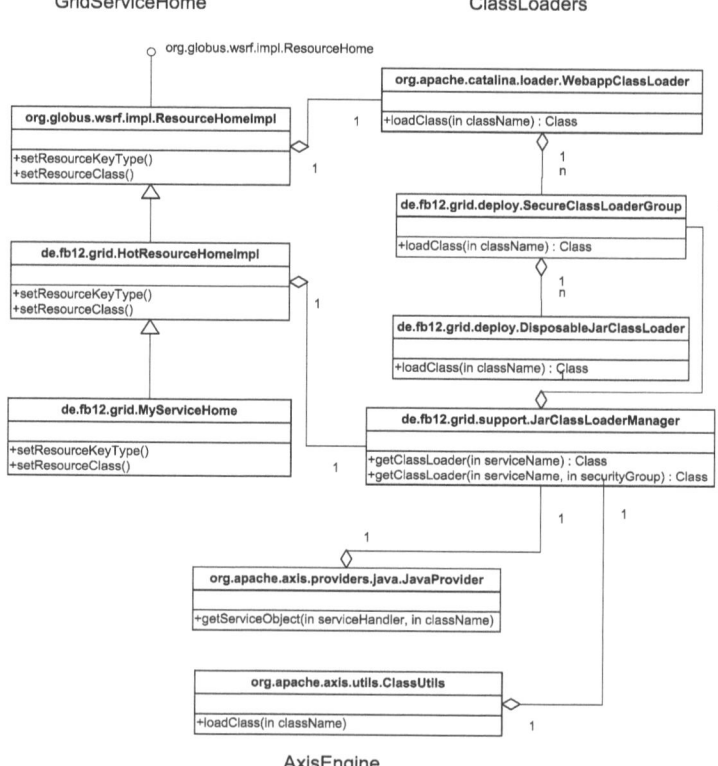

Fig. 6: *Relationship of ResourceHome implementations and class loaders*

8.5 Implementation of a Service-Oriented Ad Hoc Grid

To remove a service, the in-memory registry entries are deleted from the JNDI and Axis system registries. Once the service information has been removed, no new service instances can be created. Running instances of a service previously created are untouched by this process. To deploy a new version of a service, no explicit unloading of the old service classes is required, since the new version of the service will be created using a new ClassLoader. If in addition to the service information the service instances are to be removed, the central manager used by the JarClassLoaderManager can be used to access the ClassLoaders of the separate services to free the resources and unload the classes. Only then can the jar files be deleted, since otherwise active services might try to lazy load classes after the containing jar files have already been removed.

Fig. 7 shows a snapshot of the ClassLoader hierarchy in the system, after three different Grid services were instantiated. The GT4 environment and our hot service deployment mechanism are loaded by the Tomcat WebAppClassLoader. The remote clients call the deployment service where the HotFactoryCallBackImpl creates a DisposableClassLoader for each service creation request received. The GridServiceHandle is then returned to the clients. After the instantiation of the third service 'C', its implementation was updated and re-instantiated as 'C*'.

This central component of our ad-hoc Grid now allows services to be installed on demand on nodes running the HDS.

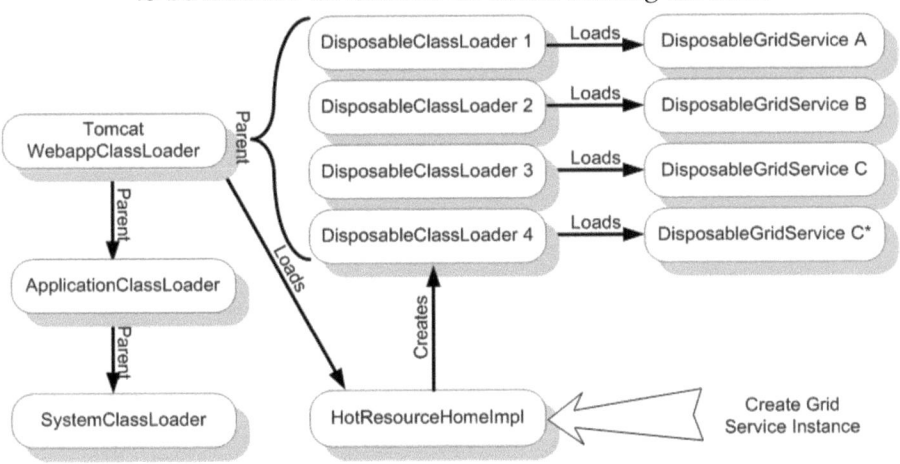

Fig. 7: ClassLoader hierarchy

8.5.6 Service Security

In this section, the security threads of an ad hoc Grid environment are presented in more detail. The discussion is restricted to Java applications running in the Grid; the security threats associated with "legacy" code are beyond the scope of this chapter and are discussed elsewhere [19]. Since the separate Grid services running on one Grid node are all hosted by Axis in GT4, they run within the same JVM and the classes are loaded by the same class loader. As a consequence, interaction between the classes is possible, thus offering malicious code the possibility to harm running services. To illustrate the problem, we introduce a target service which we will then attack. The service code is listed in Fig. 8. The main class A lies in the package de.fb12.Grid.services. It has two static methods for service internal use to access and set some data. It also has a public service method which is called by the clients wanting to use the service. The get and set methods are declared as protected methods, which restricts access to these methods to classes declared in the same package.

```
package de.fb12.grid.services;
class A
{
    static protected Data getData()
    { ... }

    static protected void setData(Data data)
    { ... }

    public void doSomething()
        throws RemoteException
    {
        //do something
    }
}
```

Fig. 8: *Target service main class*

8.5 Implementation of a Service-Oriented Ad Hoc Grid

The static get and set methods are the origin of the first intra-engine service security issue and enable an *intra-engine service data attack*. In general, an intra-engine service data attack is made possible if the service targeted for attack uses singletons, or any other method to access internal objects which does not require specific object references, like static methods. It is then possible to introduce a service which can modify these objects simply by using the same package as the target service and calling the methods on those objects. Using this attack, the internal state of an object belonging to a different service can be modified. Fig. 9 shows how data from the target service in *Fig. 8* is accessed and then replaced by a different data object.

Since the get and set methods are declared as protected, the attacking class was placed in the same package as the target class. In a standard Grid computing environment, we could configure a security manager to restrict access to the package de.fb12.grid.services and thus prevent the malicious class entering our package and accessing our target service. This is done by setting *package.definition* = *de.fb12.grid.services* and *package.access* = *de.fb12.grid.services* in the *java.security* properties file, effectively blocking creation of new classes in the protected package and access to the package from unauthorized classes. These security properties must be set before starting the system. In an ad hoc Grid environment, this is not feasible since new services in custom packages will have to be inserted during runtime and thus could not be protected. Furthermore, legitimate access to these existing services or legitimate deployment of new services in the same package should be possible. The standard approach in Java to allow this is to set specific security permissions (shown in Fig. 10) for the code base trying to access a package or create a new class in that package. In the case of GT4, which uses Tomcat as the WebApplication host, these settings are stored in catalina_home/conf/catalina.policy and are loaded during startup. Again, this is not a usable solution for ad hoc Grid computing since code bases not registered during startup might need legitimate access to certain packages. To prevent illegal access, we require a sandboxing system which allows us to protect classes from intra-engine service data attacks while at the same time allowing dynamically added services to access our classes if they have sufficient security clearance and the certificates to prove it.

8 Service-Oriented Ad Hoc Grids

```
package de.fb12.grid.services;
public class EvilService{
    doDataAttack(){
        stolenData=A.getData();
        A.setData(evilData);
    }
}
```

Fig. 9: The data attack service

```
grant codeBase
"file:\${wsrf.home}/WEB-INF/lib/MyService/my-service.jar"
{
  permission java.lang.RuntimePermission
  "defineClassInPackage.de.fb12.grid.services"
};
```

Fig. 10: Excerpt from the catalina.policy file

A different form of attack is the *intra-engine service code attack*. If the target service is loaded after the attacking services or loads classes on demand (the standard procedure in Java), it is possible to introduce foreign code into the service by pre-empting the load procedure. If, for example, the target service has a Class A which loads Class B at a certain time during its operation, a malicious service can use that as an entry point for an attack. *Fig. 11* illustrates the attack.

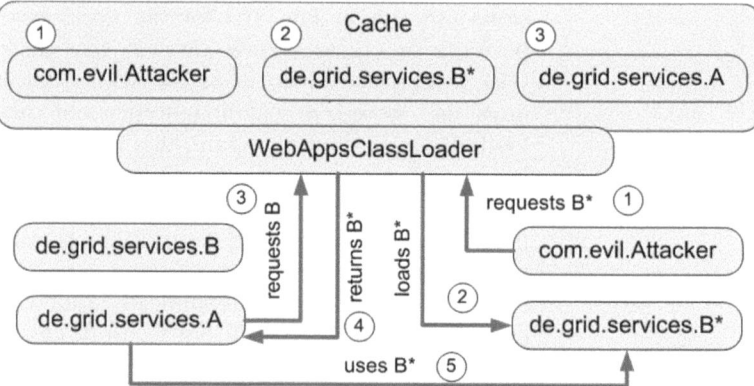

Fig. 11: Intra-engine service code attack

8.5 Implementation of a Service-Oriented Ad Hoc Grid

The attacking service defines a Class B* with the same fully qualified name and with the same method signature as Class B and then loads that class (11.1). If this is done prior to the loading of B in the target service, the malicious code has been successfully introduced into the system (11.2). When A tries to load B (11.3), the ClassLoader will see it has already loaded a class with the fully qualified name of B (namely B*) and returns the class from its cache (11.4). As a consequence, A now executes B* instead of B (11.5). This attack can even be used to replace the service A completely if A is deployed to a node where the malicious service is already running.

Fig. 12 and 13 show how this is done. Fig. 12 shows the malicious replacement A* of the target service A. It has the same method signature as the legitimate service A and thus can function as its replacement but executes the attack code instead when, for instance, the method doSomething is called.

```
package de.fb12.grid.services;
class A{
    static Data getData()
    { ... }
    static void setData(Data data)
    { ... }

    private void doSomething()
      throws RemoteException
    {
        //do something malicious
    }
}
```

Fig. 12: *Malicious replacement of the target service*

```
package com.evil;
import de.fb12.grid.services.B;
public class EvilService{
    private doCodeAttack(){
       A a = new A();
    }
}
```

Fig. 13: *The code attack service*

8 Service-Oriented Ad Hoc Grids

If the target service is deployed onto a Grid note where the malicious service A* is already loaded, the Axis class loader will simply return A* when A is supposed to be loaded, since A* has the same fully qualified name as A. Fig. 13 shows how the attacking servce loads the A* class.

If service code is protected by placing the computation into a separate class which can only be instantiated by classes of the same package, as shown in Fig. 14, the attacking service can either be placed in the same package or it can avoid instantiating the class altogether and simply load the class explicitly, as shown in Fig. 15.

```
package de.fb12.grid.services;
class B{
   protected B()
   { ... }
   private void doMore(){
      //do something good
   }
}
```

Fig. 14: Target service helper class B

```
package com.evil;
public class EvilService{
   private doCodeAttack(){
      try{
         EvilService.class.getClassLoader().
            loadClass("de.fb12.grid.services.B");
      } catch (ClassNotFoundException e){
         e.printStackTrace();
      }
   }
}
```

Fig. 15: The code attack service 2

8.5 Implementation of a Service-Oriented Ad Hoc Grid

The solution to the above security problems requires that we are able to deploy services into separate sandboxes which protect the services from illegal access.

8.5.6.1 An Approach to Intra-Engine Service Security

The attacks described above lead us to propose the following intra-engine service security requirements: A service must be able to be deployed into a private sandbox if it does not want its classes to be accessed by other services. Services wishing to form a group in which classes can be shared require a secure grouping mechanism which allows all services within the group to share classes but services outside of the group are denied access. Both mechanisms must function in an on demand fashion, i.e., normal operation of the Grid node must not be disrupted. Services already running on the system must be unaffected by the introduction of new services and new security groups.

In GT4, the intra-engine service attacks described previously are made possible by the fact that GT4 loads all Grid services within the same class loader. The basic idea of our solution to this problem is to use the ClassLoader hierarchy introduced in the previous section to enable the dynamic loading and reloading of classes to also provide intra-engine service security. In its most basic form, each Grid service is loaded within its own ClassLoader functioning as a sandbox and as such its classes and resources are private and cannot be accessed by any other service. This ensures that services using singletons cannot be hijacked by malicious services, and foreign code cannot be inserted into the program flow.

8.5.6.2 Service Sandboxing in Axis

To enable the required secure loading process, the Class-Utils and JavaProvider classes provided by Axis needed to be modified. To ensure that service classes are only loaded by our sandboxed class loader, the ClassUtils were modified to retrieve the current message context and checked whether the current loader request was triggered by a service or by the container itself. If it was triggered by a service, it checks whether it is a service which is registered with our ClassLoaderManager and should be protected. If that is the case, the class is loaded using the appropriate service ClassLoader and Axis is prevented from loading the classes in its WebAppsClassLoader and thus breaking our sand-

173

8 Service-Oriented Ad Hoc Grids

box. Otherwise, the Axis class loading is unmodified, allowing all normal operations to proceed unhindered. The second place where Axis could try and load the service classes into its own WebAppsClassLoader is the JavaProvider. Fig. 16 shows the modification to the loading process needed to protect the service classes. Similar to above, we check whether the service is registered with our framework and if that is the case we pre-empt the Axis loading mechanism and use our own ClassLoaders.

```
protected Class getServiceClass(String clsName,
    SOAPService service, MessageContext msgContext)
    throws AxisFault {
      if(JarClassLoaderManager.isRegistered(service.name)){
        ClassLoader cl = JarClassLoaderManager.
          getClassLoader(service.name)
        return cl.loadClass(clsName)
      }
      else {
        standard Axis behaviour
      }
    }
}
```

Fig. 16: *The modified getServiceClass operation in the Axis JavaProvider*

Through these modifications to Axis, inter-service communication is now confined to using web service calls and thus ensures that proper authentication between services must be observed. In many cases, this approach suffices to protect the service being deployed while still allowing unhindered operations within the service.

8.5.6.3 Secure Sandbox Groups

If it is necessary that two services be able to communicate directly using class references to create a composed web service application, they must group their ClassLoaders together using a SecureGroupClassLoader provided as part of our intra-engine service security infrastructure. A service specifies which group it wants to join either by passing the groupId to the Hot Deployment Service or by setting the parameter ** in the server-deploy WSDD of the service. The SecureGroupClassLoader responsible for the group is a

8.5 Implementation of a Service-Oriented Ad Hoc Grid

parent ClassLoader to all service ClassLoaders in that group. It enables inter-service communication in two ways: First, separate communication classes are placed in the SecureGroupClassLoader which can be accessed by all child ClassLoaders. This is the preferred way as defined by Java to allow classes in sister ClassLoaders to communicate. For instance, the interface class of an object to be used by classes in two sister ClassLoaders is placed with the parent so it can be accessed by both children. The implementing classes are placed in both child ClassLoaders and object references can be passed between ClassLoaders as long as only the interface defined in the parent ClassLoader is used. This is the traditional way to allow code-based interaction between services but it requires that the communication classes are placed in the parent ClassLoader. The disadvantage of this approach is that if the communication classes need to be replaced, all child ClassLoaders must be discarded because the SecureGroupClassLoader must be replaced. So, even if only two services use the communication classes, all services must be undeployed to update the communication classes. To avoid this problem, the SecureGroupClassLoader is capable of emulating a flat namespace for its child ClassLoaders while still allowing hot deployment and hot undeployment of component parts of the composed web service application.

When a service ClassLoader joins the SecureGroupClassLoader, the SecureGroupClassLoader checks which classes the service ClassLoader is capable of loading and stores that information internally. If a different service within the same group tries to load one of those classes, its own ClassLoader will not be able to find the class and thus asks its parent, the SecureGroupClassLoader. The SecureGroupClassLoader then checks whether one of the other service ClassLoaders can load the requested Class and passes the request on to that ClassLoader before passing the request on to its parent ClassLoader, the WebAppsClassLoader. This, of course, only works if each Class is only defined once within all ClassLoaders in the same group. If different versions of one and the same class can be accessed from the same ClassLoader, TargetInvocation and ClassCastExceptions will be the result. However, this ClassLoading mechanism was designed to allow tightly coupled web services to be composed into a web application, so it is very unlikely and undesirable that the same class will be defined in two different places, since the idea

8 Service-Oriented Ad Hoc Grids

of tight integration was to be able to reuse the classes of the other services. In the case of such class duplication, the less tightly coupled composition via service calls is the preferred way of linking different web services, and the grouping function should not be used.

8.5.6.4 Undeployment of Grouped Services

Undeployment of services is more complex if the service to be undeployed is in a group, since classes from services loaded in different ClassLoaders can have references to each other. To prevent these classes from being undeployed and crashing the system when one of the other services tries to access undeployed classes, the SecureGroupClassLoader stores the information which service ClassLoaders have interacted with each other and denies undeployment requests to these services unless all service ClassLoaders which are coupled to it are undeployed at the same time. Services loaded in the same group but which have not accessed classes of the services to be undeployed remain unaffected by this process. This is a clear benefit compared to the standard approach of placing the communication classes in the parent ClassLoader.

As an example, Fig. 17 shows four Grid services which are joined into a group by one SecureGroupClassLoader. Service A defines classes U and V where U uses W which is defined by Service B. Service C defines classes X and Y and Service D defines class Z. Class Y uses Z and Z uses X. That means, Service B cannot be undeployed while A is alive, and Services C and D can only be undeployed together.

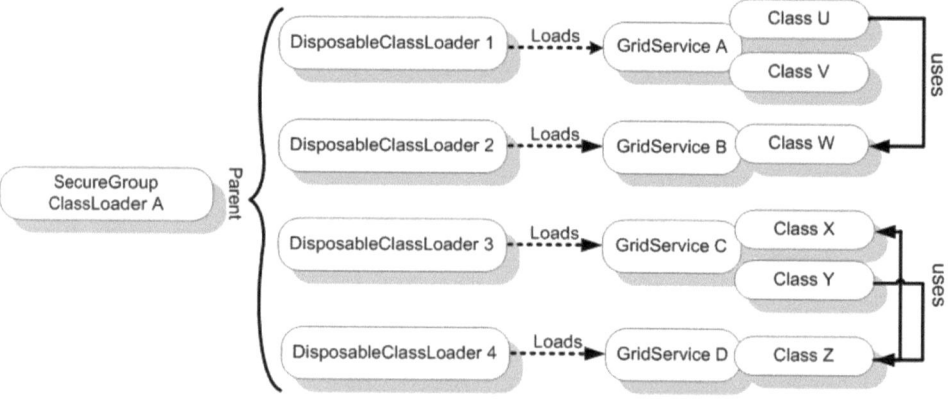

Fig. 17: ClassLoader group interaction

8.5 Implementation of a Service-Oriented Ad Hoc Grid

8.5.6.5 Group Access

To be able to securely group different service ClassLoaders together, access control to the grouping function is required. In our current implementation, when a group is created it has one owner who gets an asymmetric key pair to enable access control to the group. The private key is used by the group owner to sign Grid Service Archives (GARs) which are to be admitted to the group. The public key is used to identify the group and to check whether the GARs submitted for deployment are permitted to join the group in question. When the deploy method in the HotDeploymentService is called, the HotDeploymentService checks wether the GAR submitted for deployment was signed by the private key using the public key for that group. If the GAR was signed correctly, the deployment process is allowed and the service ClassLoader is added to the SecureGroupClassLoader of that group; if not, the deployment process is aborted and no changes to the Grid environment are made.

Fig. 18 shows a snapshot of the complete ClassLoader hierarchy in the system, after four different Grid services have been instantiated in two separate security groups. Services A through C are deployed in the same group and thus can access each others' Class definitions. Service D is deployed in its own group and thus is protected from direct code access by any of the other services deployed on this node.

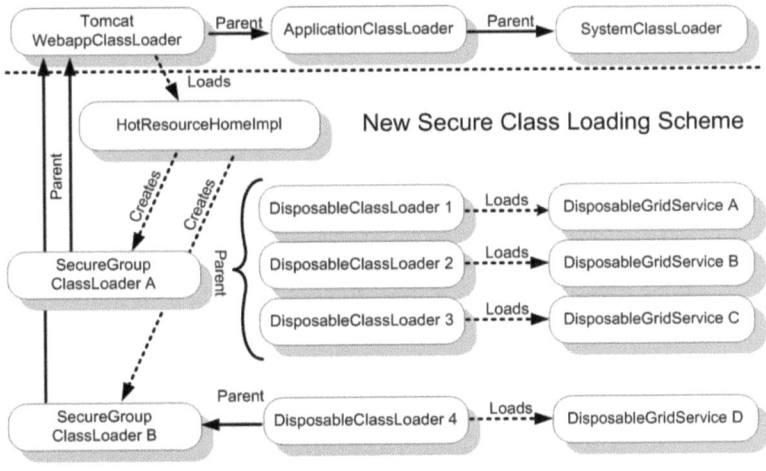

Fig. 18: Hierarchy of the ClassLoader instances

The above solution to inter-service security shows one possible way of protecting services from attacks within the same web service engine on which the service is running. Since with the progressive adoption of Grid technologies in the scientific and business communities, intra-engine inter-service security will become more relevant as more users will share Grid nodes. It would be best if the WSRF specifications deal with this topic. We propose that the requirements posed at the beginning of this section be formulated in a platform independent way, which nonetheless binds WSRF implementations to enforce intra-engine inter-service security on all platforms. The specification should then be integrated in the WSRF specifications family.

8.5.7 Service Trust

The following terminology is used in this section: A collaboration in an ad hoc Grid takes places between *interaction partners*. An interaction partner is either a *service provider* (e.g. a node to host and provide a service, or a service instance running on the provider node) or a *service consumer* (e.g. a node that requests a service from a provider (which includes the request to deploy and perform a service at the provider), or a service instance running on the consumer node). There are two major aspects that influence the selection or acceptance of an interaction partner in a service-oriented ad hoc Grid environment:

1. The identity of the interaction partner or more specifically the trust that one can put in the credibility of the identity an interaction partner claims to have.

2. The past behaviour of the interaction partner as an indicator for its future behaviour. This behaviour can be rated considering a multitude of dimensions, such as the accuracy of delivered results, actual costs compared to expected costs, availability of the service, response time, or fault and intrusion properties. Furthermore, the trust values might be different for different applications/services the interaction partner offers or requests.

In the decision processes during an interaction among consumers and providers of services, it is important to know in which context trust is considered. These contexts are, for example, the decision whether a consumer is eligible for using service instances offered by a provider or the decision whether a provider should be preferred over another provider for a specific service.

8.5 Implementation of a Service-Oriented Ad Hoc Grid

In most cases, an interaction partner is not able to judge trustworthiness based on personal and direct experiences. A socially inspired model using several dimensions of trust that builds on exchanges of experiences and recommendations is useful to decide whether to trust another interaction partner or not.

In the following, a flexible trust model and a system architecture for collecting and managing multidimensional trust values are presented. Both identity and behaviour trust of the interaction partners are considered. A proposal for establishing the first trust between interaction partners is made, and the possibility to continuously monitor the partners' behaviour trust during an interaction is provided. Our trust system can be configured to the domain specific trust requirements by the use of several separate trust profiles covering the entire lifecycle of trust establishment and management.

8.5.7.1 Trust Model

The details of our trust model are presented in [38]. In this model, the trust that interaction partners have for each other is influenced by both identity trust and behaviour trust. Although identity trust is part of the Grid authentication and authorization process, its value is nevertheless related to the overall trust value of an interaction partner. It expresses the belief that the partner is who it claims it is.

Depending on the situations, participants can have different preferences and requirements for the future interaction partners. Considering the relationship between Quality of Service (QoS) and trust [1], different QoS properties like availability of the service offered, accessibility of the service, accuracy of the response provided by the service, response time, cost of the services offered, security etc., can be considered and modelled as behaviour trust elements that a consumer uses to rate a provider. In a similar way, the total number of (concurrent) requests coming from a consumer or the size of the packets received from it can be considered as behaviour trust elements from the point of view of a provider.

Trust is a multidimensional value that can be derived from different sources when trying to determine the trust value for an interaction partner. We distinguish between three of such sources. First, there is direct and personal experience from past collaborations with a partner. A second source of information is recommendations from known sources (i.e. partners for which direct experiences exist). Finally, recommendations from other nodes/services in the Grid may be considered and then a path can be found using the known partners of known partners and so on.

To allow users to weight the different sources for the total trust differently in different situations, our model provides a profile vector of all trust sources an interaction partner A may use for the rating of another interaction partner B. Using this trust profile vector, partner A can calculate the resulting normalized trust to put into partner B. The resulting normalized trust value is only used in the decision to interact with a certain partner B. It does not affect the experience value, because this value only depends on the outcome of the subsequent interaction.

8.5.7.2 First-Trust Problem

Consider the situation when a user completely new to a Grid environment enters the network. He or she has no personal experience with any of the service providers. The usual strategies for selecting a partner to interact with do not apply in this situation. We distinguish two different basic strategies for "initializing" trust. One is a rather open approach to assign an initial trust value slightly above the minimal trust threshold to every partner, effectively giving the partner a chance to prove itself trustworthy without prior verification. We refer to this method as "warm start phase". In contrast, there might be scenarios with a higher demand on dependability in which a partner is tested by performing validation checks and deriving initial behaviour trust from these interactions. Obviously, this trust establishment phase through a "cold start" comes at a comparably high price.

The problem of establishing first trust may be seen both from a service consumer as well as a service provider point of view. We believe that a trust management environment for service-oriented ad hoc Grids must be flexible enough to allow specification of the strategy to be used in either role and on an application basis. In addition to these two basic strategies, further strategies for first trust establishment may be specified in the system.

8.5 Implementation of a Service-Oriented Ad Hoc Grid

8.5.7.3 Verification Techniques

In an ad hoc Grid environment, it might be desirable to verify that a particular partner stands up to an assumed or previously offered behaviour. The extent to which verification is performed may vary depending on application scenarios or various user requirements. Also, the need for verification of the partners' behaviour may arise in both roles (i.e. consumer or provider) of a service consumption scenario. Partners will continuously monitor the interaction process among each other, and in case of discovered anomalies in the behaviour of the other, the consumers and/or providers will re-organize their scheduling or access policies accordingly.

The different aspects of the partners' behaviour (e.g. availability, response time, accuracy, etc.) are criteria for developing verification strategies. In the following, we will only consider the accuracy of the responses coming from a service provider as an example and refer to this dimension as behaviour trust for brevity (note, however, that this is only one dimension of behaviour trust to be considered between different partners).

The strategy to use for the verification of the accuracy of responses to be expected from one provider may vary depending on certain constraints such as the additional acceptable cost for performing the verification operations. The following verification strategies might be applied:

1. *Challenge with known tasks* - A service consumer may prepare a particular test challenge for which it knows the correct result. In this case, the consumer can directly verify if a service provider was able to calculate the correct response for this challenge.

2. *Best of n replies* - A more feasible verification technique is similar to the one that is used by SETI@HOME [30]. The validity of the computed results is checked by letting different entities work on the same data unit. At the end, a majority vote establishes the correct response.

3. *Human in the loop* - In some applications, it might be impossible to construct automatic verification or result rating modules. In such cases, it can still be helpful to involve human users in the process of verification. This technique relies on presenting the results to the user, offering the abil-

ity to directly express a value to be taken into the trust calculation.

In our approach, it is possible for each of the partners to develop their personalized trust preferences towards the interaction partners. These preferences include the initialization values that the user is willing to assign to each of the new partners, the selection of sources for getting trust information from (recommendations), the interaction partners the participant collaborates with and verification strategies for all the trust elements. The consumer may choose between verifying the accuracy of every single answer coming from the provider ("trust no one") or to verify the accuracy of only a part of the responses coming from the provider ("optimistic trust"). In order to minimize added costs, we propose to couple the frequency of this partial verification technique with the behaviour trust associated with a particular partner in the environment. From the consumer side this means that for a non-trusted provider every single response is verified and for a fully trusted provider only a minimum of the responses coming from that specific provider has to be verified. The result of the verification operations will directly be used to alter the behaviour trust regarding accuracy.

8.5.7.4 System Architecture

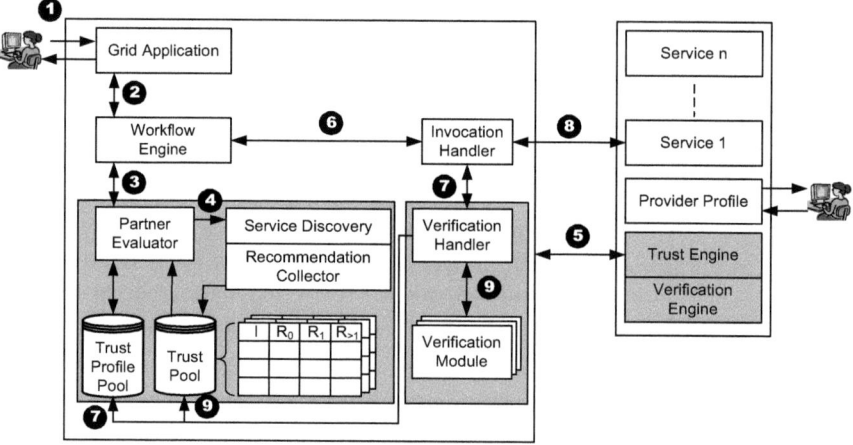

Fig. 19: Architecture of a grid system supporting our trust model.

8.5 Implementation of a Service-Oriented Ad Hoc Grid

A system architecture supporting trust management in service-oriented Grid applications is presented in Fig. 19. The system consists of two main components, the *trust engine* and the *verification engine*. The trust engine manages trust values and offers partner discovery and rating functionality to higher level applications, such as workflow engines or job scheduling systems. The verification engine handles the verification of Grid service results and generates the necessary feedback for the trust engine regarding the partner. For brevity, we will focus our discussion on the service consumer use of those platform components.

The user starts with specifying his or her trust requirements along with the input data to a trust enabled Grid application (1), which in turn uses the workflow engine of the local service-oriented Grid platform (2). To enable the selection of trusted services, the decision is made based on a rated list of potential partner services that is obtained from the trust engine (3). The trust engine uses its service discovery component to discover individual services (4) and to collect recommendations from other trust engines (5). These values are stored in the local trust pool to be used in subsequent interactions. The user specified trust profile is also stored in a trust pool for later reference and use by other components in the trust engine. The information gathered by the trust engine is now processed according to the user's trust profile specification and passed on to the workflow engine which then can use the partner services according to the rating generated by the trust engine.

Invocation of external services is then delegated to an invocation handler (6). The invocation handler consults the verification engine (7) to determine whether a call has to be replicated or redirected (e.g. to perform the best of n verification strategy). The verification engine considers the trust profile managed by the trust engine (7), allowing, for example, cost-trust-ratio relations to be taken into account. The resulting invocation is carried out at the selected partner services and results - both synchronous and asynchronous (notification) results - are then collected by the invocation handler (8) and verified through the verification engine, using a strategy and verification module consistent with the user supplied trust profile (9). The overall result of this process is then passed to the workflow engine that collects results for the application to present them to the end user.

8 Service-Oriented Ad Hoc Grids

The configuration of the trust engine by use of trust requirement profiles influences three phases during execution of an application workflow. These main phases are addressed by the three arrows in Fig. 20.

The initialization profile determines the influence and scope of factors used for the initialization of trust values to be used in an interaction. It allows to manually assign trust values to certain interaction partners, as well as specifying how trust recommendations of partners are handled and weighted. This profile specifies the behaviour of the local platform in a situation that requires the establishment of first trust.

The source selection profile determines the selection of behaviour trust dimensions (e.g. availability, accuracy) as well as trust sources (e.g. personal experience, recommendations from directly known partners) to determine a partner ranking according to the application needs. This allows a user to take accuracy trust recommendations from known partners into account with a higher weight than, for example, availability values (which might be caused by the different network locations) coming from the same partner.

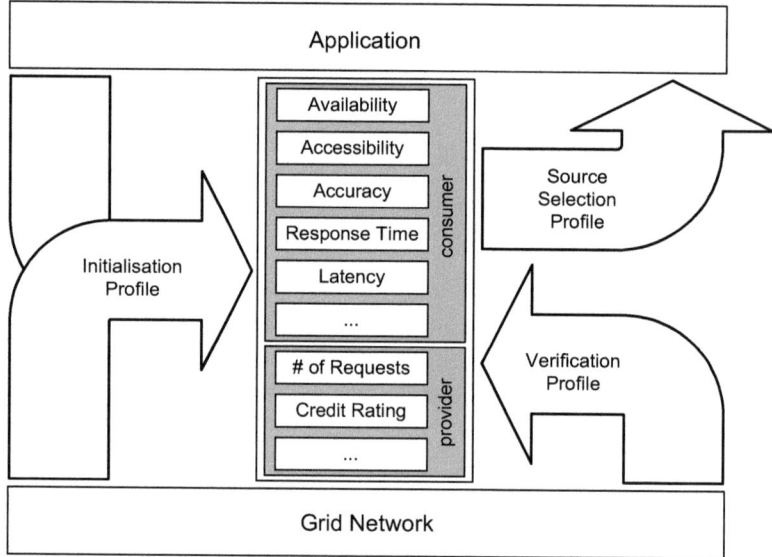

Fig. 20: *Trust profile elements influencing the stored trust values as well as application decisions.*

The verification profile specifies which verification strategies are to be applied to the results of partner service invocations and the feedback parameters into the trust engine. In this profile, the user specifies how breaches of assumed service level agreements should influence the future interactions with a partner since they are fed back into the trust store for this application and partner service. This profile also dynamically determines the frequency of verification to allow a fine grained control over costs incurred by result verification.

The user may reference custom verification module implementations that are instantiated as a plug-in to the verification engine in order to allow a high flexibility and result content specific verification. Similarly, the verification frequency strategy may be a simple one supplied with the trust engine (e.g. verify every n-th result) or a custom strategy implementation supplied as a plug-in by the application developer (e.g. start with a high verification frequency, back off after successful interactions, fall back to frequent verification after a failed verification). Verification can be handled in a trust dimension specific manner; thereby it is possible to penalize a partner more severely for failed accuracy verification than for unavailability.

8.6 Conclusions

In this chapter, we have presented a service-oriented ad hoc Grid as a spontaneous fusion of cooperating heterogeneous computing nodes into a logical community without a preconfigured fixed infrastructure and with only minimal administrative requirements. The main goal of an ad hoc Grid is to provide computing resources on demand to every participant. We have discussed the main issues involved in building such an ad hoc Grid. Approaches to solve the problems of node and service discovery, hot service deployment administration, inter-service security and trust management were presented.

There are several areas for future research. First, the current prototypical implementation should be evaluated by testing its behaviour and performance with real-world Grid applications. Second, several management wizards for deployment, monitoring and updating need to be developed for increasing the ease of use of an ad hoc Grid. Third, the security concerns of "legacy" code written in C, C++ or Fortran must be further addressed, based on a first proposal presented in [19]. Fourth, the integration of suitable workflow engines into the Grid application development process should be investigated. Fourth, the issues of licensing, accounting and billing are important in any Grid environment and thus should be studied particularly for the ad hoc Grid scenario. Finally, autonomic computing principles to increase the robustness and self-management properties of an ad hoc Grid are interesting areas for future work.

8.7 Acknowledgements

This work is financially supported by Siemens AG, Corporate Technology, München, by an IBM Faculty Award (Eclipse Innovation Grant) and by the Deutsche Forschungsgemeinschaft (SFB/FK 615, Teilprojekt MT).

8.8 References

[1] A. S. Ali, O. Rana and D. W. Walker, "WS-QoC: Measuring Quality of Service Compliance", *In Proceeding of the Second International Conference on Service-Oriented Computing, Short Papers (ICSOC)*, New York, USA, 2004, pp. 16–25.

[2] K. Amin, G. von Laszewski, and A. R. Mikler, "Toward an Architecture for ad hoc Grids," in *International Conference on Advanced Computing and Communications, India*, 2004.

[3] N. Andrade, L. Costa, G. Germoglio, and W. Cirne, "Peer-to-peer Grid Computing with the Ourgrid Community," in *Proceedings of the 23rd Brazilian Symposium on Computer Networks*, 2005 (to appear).

[4] N. Andraden, W. Cirne, F. Brasileiro, and P. Roisenberg, "Ourgrid: An Approach to Easily Assemble Grids with Equitable Resource Sharing," in *Proceedings of the 9th Workshop on Job Scheduling Strategies for Parallel Processing*, 2003, pp. 61 – 86.

[5] "Apache Tomcat 5.0"
http://jakarta.apache.org/tomcat/index.html.

[6] F. Azzedin and M. Maheswaran, "Evolving and Managing Trust in Grid Computing Systems." in *Proceedings of the IEEE Canadian Conference on Electrical and Computer Engineering*, Winnipeg, Canada, 2002, pp. 1424-1429.

8.8 References

[7] F. Berman, G. Fox, and T. Hey, *Grid Computing: Making the Global Infrastructure a Reality.* Wiley, 2003.

[8] E. Bertino, P. Mazzoleni, B. Crispo, and S. Sivasubramanian, "Towards Supporting Fine-Grained Access Control for Grid Resources," in *10th IEEE International Workshop on Future Trends of Distributed Computing Systems,* 2004, pp. 59 – 65.

[9] A. Chakravarti, G. Baumgartner, and M. Lauria, "The Organic Grid: Self-Organizing Computation on a Peer-to-Peer Network," in *IEEE Transactions on Systems, Man and Cybernetics, Part A,* 2005, pp. 373– 384.

[10] Y. Chawathe, S. Ratnasamy, L. Breslau, N. Lanham, and S. Shenker, "Making Gnutella-Like P2P Systems Scalable," in *Proc. of the 2003 Conference on Applications, Technologies, Architectures, and Protocols for Computer Communications.* ACM Press, 2003, pp. 407–418.

[11] E. Christensen, F. Curbera, G. Meredith, and S. Weerawarana, Web Services Description Language, 2001, http://www.w3.org/TR/wsdl.

[12] EGEE, "Enabling Grids for eScience in Europe: Executive Summary," 2004, http://egee-intranet.web.cern.ch/egee-intranet.

[13] I. Foster and A. Iamnitchi, "On Death, Taxes, and the Convergence of Peer-to-Peer and Grid Computing," in *2nd International Workshop on Peer-to-Peer Systems (IPTPS'03),* 2003, pp. 118–128.

[14] I. Foster and C. Kesselman, *The Grid 2: Blueprint for a New Computing Infrastructure.* Morgan Kaufmann, 2003.

[15] I. Foster, D. Berry, A. Djaoui, A. Grimshaw, B. Horn, H. Kishimoto, F. Maciel, A. Savvy, F. Siebenlist, R. Subramaniam, J. Treadwell, and J. Von Reich. "The Open Grid Services Architecture", Version 1.0, Whitepaper GGF, 2004, pp. 1-19.

[16] I. Foster, C. Kesselman, J. Nick, and S. Tuecke, "The Physiology of the Grid: An Open Grid Services Architecture for Distributed Systems Integration," in *Open Grid Service Infrastructure WG, Global Grid Forum,* 2002, pp. 1-31.

[17] T. Friese, B. Freisleben, S. Rusitschka and A. Southall, "A Framework for Resource Management in Peer-to-Peer Net-

works" in *Proceedings of the Int. Conference NetObjectDays 2002 LNCS 2591, Erfurt, Germany*, Springer, 2002, pp. 4-21.

[18] T. Friese, M. Smith, and B. Freisleben, "Hot Service Deployment in an Ad Hoc Grid Environment," in *Proceedings of the 2nd Int. Conference on Service-Oriented Computing, New York, USA*. ACM Press, 2004, pp. 75–83.

[19] T. Friese, M. Smith, and B. Freisleben, "Native Code Security for Java Grid Services", submitted for publication, 2005.

[20] D. Gannon, R. Ananthakrishnan, S. Krishnan, M. Govindaraju, L. Ramakrishnan, and A. Slominski, "Grid Web Services and Application Factories," in *Grid Computing: Making the Global Infrastructure a Reality*, F. Berman, G. Fox, and T. Hey, Eds. Wiley, 2003.

[21] C. Germain, V. Breton, P. Clarysse, Y. Gaudeau, T. Glatard, E. Jeannot, Y. Legré, C. Loomis, J. Montagnat, J.-M. Moureaux, A. Osorio, X. Pennec, and R. Texier., "Grid-Enabling Medical Image Analysis," in *IEEE Procedings CCGrid 2005 Bio-Grid Workshop*, 2005 (to appear).

[22] T. Glatard, J. Montagnat, and X. Pennec, "Grid-Enabled Workflows for Data Intensive Applications," in *Computer Based Medical Systems*, 2005 (to appear).

[23] "The Globus Toolkit 4.0," http://www.globus.org/toolkit/

[24] N. S. Good and A. Krekelberg, "Usability and Privacy: A Study of Kazaa P2P File-Sharing," in *Proceedings of the Conference on Human Factors in Computing Systems*. ACM Press, 2003, pp. 137–144.

[25] J. Han and D. Park, "A Lightweight Personal Grid Using a Supernode Network," in *Third International IEEE Conference on Peer-to-Peer Computing*, 2003, pp. 168 – 175.

[26] A. Iamnitchi and I. Foster, "On Fully Decentralized Resource Discovery in Grid Environments," in *International Workshop on Grid Computing*, 2001, pp. 51–62.

[27] A. Iamnitchi, I. Foster, and D. Nurmi, "A Peer-to-Peer Approach to Resource Discovery in Grid Environments," in *Proc. of the 11th Symposium on High Performance Distributed Computing*, 2002, p. 419.

8.8 References

[28] "IBM Making a Commitment to the Next Phase of the Internet," http://marianne.in2p3.fr/dataGrid/documents/IbmGrid.pdf.

[29] Introduction to UDDI Important Features and Functional Concepts, Technical White Paper, pages 1-11, 2004, http://uddi.org/pubs/uddi-tech-wp.pdf

[30] E. Korpela, D. Werthimer, D. Anderson, J. Cobb, and M. Lebofsky, "SETI64home - Massively Distributed Computing for SETI," *Computing in Science and Engineering*, vol. 3, no. 1, p. 79, 2001.

[31] G. von Laszewski, I. Foster, J. Gawor, and P. Lane, "A Java Commodity Grid Kit," in *Concurrency and Computation: Practice and Experience*, vol. 13, no. 8-9, 2001, pp. 643–662.

[32] "Limewire," http://www.limewire.org/

[33] D. Martin, M. Paolucci, S. McIlraith, M. Burstein, D. McDermott, D. McGuinness, B. Parsia, T. Payne, M. Sabou, M. Solanki, N. Srinivasan, K. Sycara, "Bringing Semantics to Web Services: The OWL-S Approach", *In Proc. 1st Int. Workshop on Semantic Web Services and Web Process Composition LNCS 3387*, San Diego, USA, Springer, 2004, pp. 26-42.

[34] M. Murshed and R. Buyya, "Using the GridSim Toolkit for Enabling Grid Computing Education," in *Proc. of the Int. Conf. on Communication Networks and Distributed Systems Modeling and Simulation*, 2002. pp. 18-24

[35] V. Naik, S. Sivasubramanian, D. Bantz, and S. Krishnan;, "Harmony: a Desktop Grid for Delivering Enterprise Computations" in *Proceedings. Fourth International Workshop on Grid Computing*, 2003. pp 25 - 33

[36] NASA Advanced Supercomputing Division, "NASA Information Power Grid," 2002, http://www.nas.nasa.gov/About/IPG/ipg.html.

[37] OASIS. Web Services Resource Framework, 2004. http://www.oasisopen.org/committees/tc_home.php?wg_abbrev=wsrf

[38] E. Papalilo, T. Friese, M. Smith, and B. Freisleben, "Trust Shaping: Adapting Trust Establishment and Management to Application Requirements in a Service-Oriented Grid Environment", submitted for publication, 2005.

[39] *Project JXTA v2.0: Java Programmer's Guide*, 2003.

[40] M. Roussopoulos, M. Baker, D. Rosenthal, T. Guili, P. Maniatis and J. Mogul, "2 P2P or not 2 P2P?", *In: Proc. 3rd Int. Workshop on P2P Systems*, San Diego/USA, Springer, 2004, pp. 33-44.

[41] T. Rusitschka and A. Southall, "The Resource Management Framework: A System for Managing Metadata in Decentralized Networks Using Peer-to-Peer Technology" in *Proceedings of the first Int. Workshop on Agents and Peer-to-Peer Computing Bologna, Italy*, Springer, 2002, pp. 144-149.

[42] S. Shivle, H. Siegel, A. Maciejewski, T. Banka, K. Chindam, S. Dussinger, A. Kutruff, P. Penumarthy, P. Pichumani, P. Satyasekaran, D. Sendek, J. Sousa, J. Sridharan, P. Sugavanam, and J. Velazco, "Mapping of subtasks with multiple versions in a heterogeneous ad hoc grid environment," in *Algorithms, Models and Tools for Parallel Computing on Heterogeneous Networks*, 2004, pp. 380 – 387.

[43] M. Smith, T. Friese, and B. Freisleben. Towards a Service-Oriented Ad Hoc Grid. In: Proc. 3rd Int. Symp. on Par. and Distrib. Comp., Cork/Ireland, IEEE Press, 2004, pp. 201-209.

[44] M. Smith, T. Friese, B. Freisleben: Intra-Engine Service Security for Grids Based on WSRF. In: Proceedings of Cluster Computing and Grid, Cardiff, IEEE Press, 2005 (to appear).

[45] H. De Sterck, R. Markel, and R. Knight, "A Lightweight, Scalable Grid Computing Framework for Parallel Bioinformatics Applications", in Proceedings of the 19th International Symposium on *High Performance Computing Systems and Applications*, IEEE Press, 2005, pp. 251 – 257.

[46] I. Stoica, R. Morris, D. Karger, M. F. Kaashoek, and H. Balakrishnan, "Chord: A Scalable Peer-to-Peer Lookup Service for Internet Applications", *In Proceedings of the ACM SIGCOMM 2001 Technical Conference*, San Diego, USA, ACM, 2004, pp. 149-160.

[47] U.S. Dept. of Energy Office of Science, "Dept. of Energy and IBM Science Grid," 2001, http://doescienceGrid.org.

[48] Wikipedia Entry "Gnutella", http://en.wikipedia.org/wiki/Gnutella.

[49] The World Wide Web Consortium, "Simple Object Access Protocol (SOAP)," 2003, http://www.w3.org/TR/soap/.

9 Model Driven Development of Service-Oriented Grid Applications

M. Smith, T. Friese and B. Freisleben

9.1 Introduction

The service-oriented architecture (SOA) approach and the corresponding web service standards such as WSDL [5] and SOAP [21] are currently adopted in various fields of distributed application development (e.g. enterprise application integration, web application development, inter-organizational workflow collaboration){XE "inter-organizational workflow collaboration"}. The Open Grid Services Architecture (OGSA) [9, 8] incorporates the web service paradigm in the field of *Grid Computing* [9, 20] as an approach towards defining the *service-oriented Grid*. The service-oriented Grid paradigm offers the potential to provide a fine grained virtualization of the available resources to significantly increase the versatility of a Grid. The OGSA effort to add stateful interaction to the web service environment has also been recognized by other web service users not focused on Grid computing. As a result, the specifications of the Web Service Resource Framework (WSRF) [14] have emerged.

The WSRF introduces the notion of a *web service resource* (WS-Resource) that is formed by the combination of a *resource document* and a corresponding *web service*. It is the purpose of the resource document to capture state information for a WS-Resource while the corresponding web service remains stateless. In this way, a multitude of WS-Resources can be created using a single stateless web service implementation which captures the state of execution in multiple resource documents. The WSRF further defines web service interfaces to inspect and alter the information contained within a resource document and to receive and subscribe to notifications on property changes.

Notwithstanding its potential benefits, service-oriented Grid application development based on WSRF is a complex issue and the "learning curve" for newcomers is quite high (as experienced by many of our students). In [22], Dave Thomas argues that the complexity of modern software systems is rapidly growing because there is "too much stuff" leading to a situation where "things are so complex you need a M.Sc. to program crud!" This bold statement is particularly true for the development of service-oriented Grid software. In its basic form, the typical service-oriented Grid middleware consists of the Globus Toolkit 4.0 (GT4), Tomcat 5.5 and Axis, three large scale software projects encompassing thousands of Java classes each, not to mention half a dozen third party support libraries for GT4 alone, which an application developer needs to keep in mind. To make matters worse, currently middleware and business code are tightly coupled, requiring both business and middleware developers to have expert knowledge in both areas.

Model driven architecture (MDA) [12, 2] has been proposed as an approach to deal with complex software systems by splitting the development process into three separate model layers and automatically transforming models from one layer into the other:

1. The Platform Independent Model (PIM) layer holds a high level representation of the entire system without committing to any specific operating system, middleware or programming language. The PIM provides a formal definition of an application's functionality without burdening the user with too much detail.

9.1 Introduction

2. The Platform Specific Model (PSM) layer holds a representation of the software specific to a certain target platform such as J2EE, Corba or in our case the service-oriented Grid middleware.
3. The Code Layer consists of the actual source code and supporting files which can be compiled into a working piece of software. In this layer, every part of the system is completely specified.

MDA theory states that a PIM is specified and automatically transformed into a PSM and then into actual code, thus making system design much easier. The trick, of course, lies in the development of generic transformers capable of generating the PSM and code layers from the PIM [7, 23].

Service-oriented Grid computing is a relatively young field of distributed computing and is currently lacking any form of tool support for a model driven approach to software development. This is unfortunate since we believe that due to its high complexity and the high rate of *churn* [6] in the software technology market, a MDA approach is vital to the adoption of this new technology.

Only if business logic developers can more or less effortlessly integrate a new middleware into their system, will a widespread adoption be possible. Furthermore, the developers responsible for the integration of the middleware into the overall business system should be able to concentrate on middleware concerns and not have to cope with the business logic as well. This separation of concerns can be greatly facilitated by an appropriate MDA approach.

In this chapter, an approach to the model driven development of service-oriented Grid applications is presented. The goal of the approach is to minimize the necessary human interaction required to transform a PIM into a PSM and a PSM into code for a service-oriented Grid environment, in order to avoid that MDA becomes yet another part of "too much stuff". To further separate the Grid specific components of the PSM from the business specific components of the PSM, a UML Grid Profile is introduced and a separation of the PSM layer into two parts is proposed which make the automated transformations from PIM to PSM to code easier to implement and more transparent for system designers, developers, and users. The separation of concerns introduced on the PSM layer is mirrored on the code layer by the use of Java annotations, allowing the same business code to run in different domains simply by exchanging the annotations and thus decoupling application code and service-oriented Grid middleware.

The chapter is organized as follows. Section 9.2 discusses related work. Section 9.3 shows an example Grid application on the PIM, PSM and code layers. Section 9.4 introduces our Grid Profile, the MDA and Java annotations based approach to separation of concerns and Grid transformations. Section 9.5 concludes the chapter and outlines areas for future research.

9.2 Related Work

9.2.1 Service-Oriented Grid Computing

The Open Grid Services Architecture (OGSA) has been accepted as the foundation for service-oriented Grid computing. While OGSA describes the higher level architectural aspects of service-oriented Grid computing, the Web Service Resource Framework (WSRF) is a fine grained specification of the infrastructure required to implement the OGSA model. Several implementations of WSRF are being developed concurrently, including the Globus Toolkit 4 (GT4) [16] and WSRF.NET [3].

9.2.2 Model Driven Architecture

9.2.2.1 Grid Specific MDA

In [11], Gokhale et al. describe a Model Integrated Computing Tool called CoSMIC which deals with resource reservation and component deployment for their Grid middleware Grit. Unfortunately, Grit is a CORBA based middleware which was not designed for service-oriented Grid use, is not OGSA compliant and as such does not have much impact on our work. Furthermore, the paper does not deal with the transformation of PIM models to PSM models or separation of concerns.

In [1], measurement theory and the Logic Scoring of Preferences method are used to select Grid services. The authors describe a formal model for satisfaction based service evaluation which can be used in MDA based Grid computing. The paper does not deal with the transformation of PIM models to PSM models or separation of concerns.

To the best of our knowledge, there is no related work on MDA transformations for Grid services.

9.2.2.2 Service-Orientation Specific MDA

Since service-oriented Grid computing is an extension of the service-oriented computing paradigm, there are several relevant papers dealing with applying MDA in a web service environment. They can be categorized into two opposing approaches: A document/WSDL centric approach and a programming/UML centric approach.

A WSDL centric approach is presented in [13]. The paper discusses which components of a service belong to which layer of the MDA approach. Definitions, Operations, Port type(s), Messages, Parts and Part type(s) are placed in the PIM layer. Service, Ports and Binding(s) are placed in the PSM layer. We do not agree with this mapping since most of the components placed in the PIM are specific to a service-oriented approach and should therefore be placed in the PSM layer. The paper goes on to suggest that a document centric view is better suited to SOA models than an UML centric view.

Three WSDL centric transformations are introduced in [4]. The most relevant one for our work is the mapping from a UML based PSM to web service code. For the transformation to be applied, a full specification of the WSDL document must be created in the PSM by the developer.

In a similar way, [15] describes a mapping from EDOC to web services using a detailed user created model of the WSDL document.

None of the above approaches deal with the critical transformation from a PIM to the UML description of the WSDL document or with the separation of concerns on the PSM or code level.

A WSDL free approach is presented in [17]. A WSDL independent UML model is proposed because it offers a much clearer view of the system functionality. Automatic transformations from UML to WSDL are used to create the actual WSDL document. Since the UML model is completely free of any WSDL specific components, the developer is free to concentrate on actual business concerns. The downside is that an integral part of service-oriented systems is no longer visible in the MDA models and thus outside of the development scope.

We believe that a hybrid approach between the document/WSDL centric approach and a programming/UML centric approach is preferable. On the one hand, a clear view of the system functionality is required and on the other hand all essential components of the software system should be contained in the MDA approach.

[18, 10, 19] deal with service-oriented architectures and MDA in general, but do not address the problem of transforming PIM to a SOA specific PSM or the separation of concerns in the PSM or code layer.

9.3 MDA Meets the Grid: An Application Example

9.3.1 PIM Layer: Business View

To illustrate the issues involved in the model driven development of service-oriented Grid applications, a simple example is used throughout the rest of this paper. Fig. 1 shows the PIM view of two classes used to break a Unix password. The CryptBreaker class has two methods: one which does a brute force attack using the whole range of possible password values, the other does a brute force attack using only part of the range of values. This second method is used by the CryptClient to break a password on a number of different remote nodes each containing one instance of the CryptBreaker. The CryptBreaker class also has two attributes which store previous results and the percentage of the current brute force attack which has been completed. For the sake of clarity, the actual crypt breaking logic will not be modelled, as it is not the focus of this work. This simple PIM view contains no Grid specific components and thus offers a clear view of the business concerns. The classes depicted here are not capable of being called remotely. To enable the distributed breaking of the password, the PIM will now be transformed into a Grid specific PSM.

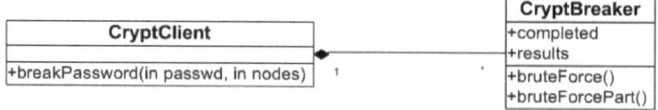

Fig. 1: Unix crypt breaker PIM

9.3 MDA Meets the Grid: An Application Example

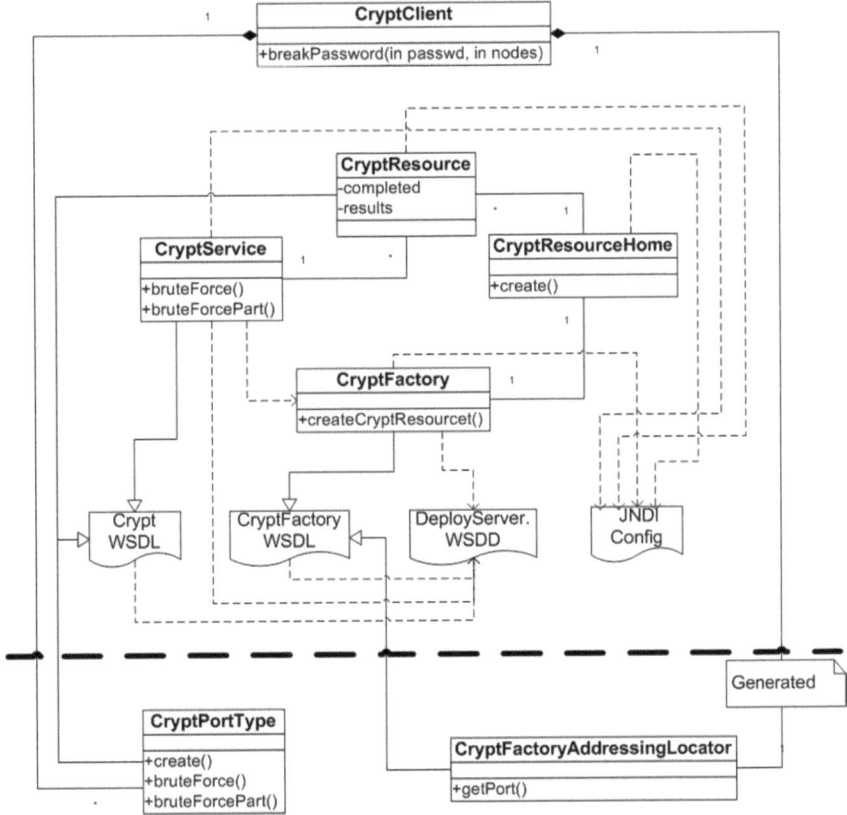

Fig. 2: Simplified Grid service UML

9.3.2 PSM Layer: Grid Service Design

Grid services in GT4 are separated into three classes: a service resource class, a service home class and the service implementation itself. The service home class is used to load resources attached to a service and the service class itself. It is recommended that each service is accompanied by a factory service used to create the service itself. Furthermore, there are four supporting files needed to deploy the Grid service: a deployment descriptor, a WSDL description of the service, a WSDL description of the factory and a JNDI configuration file used to locate the components within the container. Fig. 2 gives a simplified overview of the classes and documents comprising the Grid service. The two methods of the CryptBreaker class are placed in the CryptService

9 Model Driven Development of Service-Oriented Grid Applications

class and the attributes are placed in the resource class. The CryptClient does not work directly with the service but with its CryptPortType which it receives from the CryptAddressingLocator, two classes generated by the web service tools used in GT4.

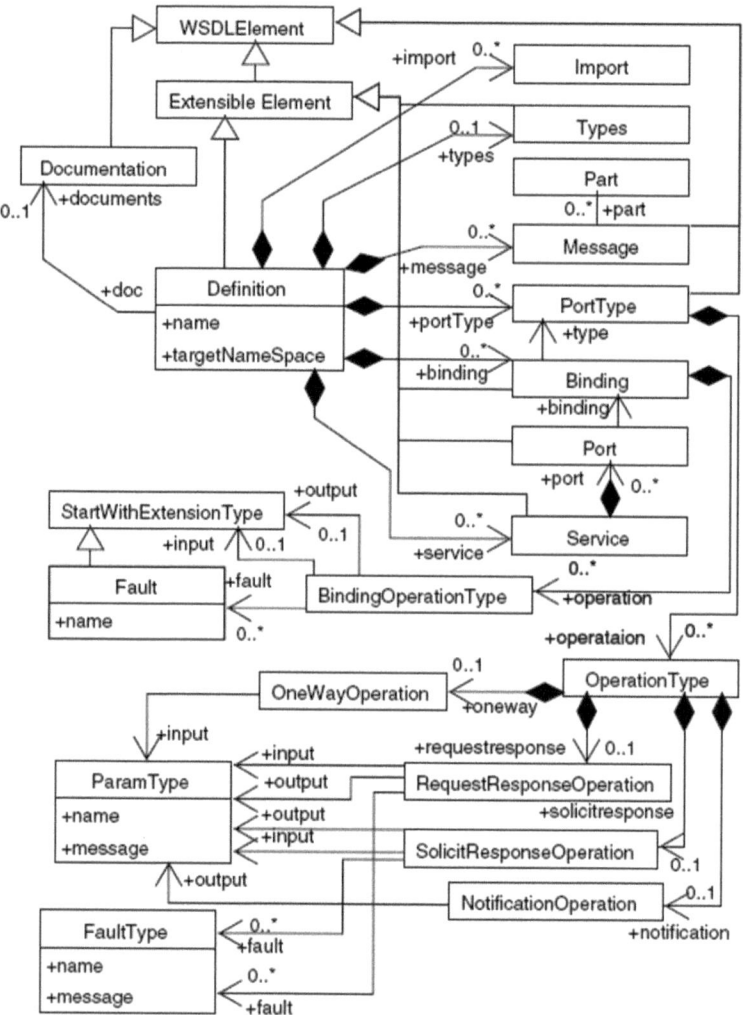

Fig. 3: *UML model of WSDL (from [4])*

The illustration is a simplified view of the PSM. To give a more complete description of the Grid service, the support files should also be modeled completely in UML. Fig. 3 shows a WSDL meta

198

9.3 MDA Meets the Grid: An Application Example

model which can be used to represent the web service part of the WSDL files. Since elements of the WSDL model and the service classes have fine grained interdependencies, this expanded model reaches a complexity which defies an easy understanding. To make matters worse, the "business code" is contained in both the service class and the resource class, so the interdependencies directly affect the business logic developers. Furthermore, the CryptClient must work with web service specific, generated classes, which forces the business logic developer to master both domains. Everything shown in the figure except for the CryptPortType and the CryptAddressingLocator must currently be created by hand both on the PSM and on the code layer.

9.3.3 Code Layer: Grid Service Implementation

To illustrate the interweaving of Grid and business concerns at the code level, we will show an excerpt of the bruteForcePart method (see Fig. 4).

```
public BruteForceResponse bruteForcePart(
    BruteForcePart complexType)
        throws RemoteException
{
String pwd = complexType.getEncryptedPassword();
int[] upperRange = complexType.getUpperRange();
int[] lowerRange = complexType.getLowerRange();
long runLength = calcRunLength(lowerRange,upperRange);
boolean decrypted=false;
while not decrypted do for each parameter in range
{
  //business logic for decrypting password
  CryptResource cryptResource = getResource();
  cryptResource.setCompleted
     ((cryptResource.getCompleted() + 1) / runLength);
}
```

Fig. 4: Crypt breaker service code

The parameter of the method is a generated Axis specific complexType which contains the password to be broken and the range of parameters to be tried. To set the completed attribute, the service needs to retrieve the resource object from the Grid container, read the value and set the new value using Grid resource specific methods. The return type must also be wrapped in an Axis generated object.

The more complex the business logic and its resource gets, the less readable the service code becomes. Fig. 5 shows what needs to be done to call the bruteForcePart method in a remote instance of the CryptService. Only the last method call is business specific, all the rest belongs to the Grid domain. The tight entanglement of business code and Grid code forces both Grid and business developers to deal with difficult code which spans at least two domains.

```
String instanceURI =
"http://xxx.xxx.xxx.xxx:8080/wsrf/" +
   "services/UnixCryptBreakerImplService";

String factoryURI = "http://xxx.xxx.xxx.xxx:8080/wsrf/"
   +"services/CryptFactoryService";

CryptFactoryAddressingLocator factoryLocator =
   new CryptFactoryServiceAddressingLocator();
EndpointReferenceType factoryEPR, instanceEPR;
CryptAddressingLocator locator =
   new CryptBreakerAddressingLocator();
factoryEPR = new EndpointReferenceType();
factoryEPR.setAddress(new Address(factoryURI));
factoryPort = factoryLocator
   .getCryptFactoryPortTypePort(factoryEPR);
CreateResourceResponse createResponse = factoryPort
   .createResource(new CreateResource());
instanceEPR = createResponse.getEndpointReference();
CryptPortType port = locator
   .getCryptPortTypePort(instanceEPR);
port.bruteForcePart("somePassword", lowerRange,
   upperRange);
```

Fig. 5: Client code to call the CryptService

9.3.4 Separation of Concerns

As shown above, the business logic is implemented in the service class and the business data (or state) is placed in the resource class. This is necessary to keep the service class stateless and web service compliant. The downside of this separation is that business concerns are spread into different Grid service specific classes, and the development domains overlap. Both on the PSM layer and on the code layer we require a separation of development concerns to offer cleaner views to each domain expert and to make the development of MDA tools for Grid services easier to implement and maintain.

9.4 An MDA Approach to Service-Oriented Grid Computing

An MDA approach to service-oriented Grid computing must deal with all three layers of the MDA model: PIM, PSM and code. The PIM layer is free of Grid concerns and thus does not need to be dealt with directly. Only the transformation logic from PIM (Fig. 1) to PSM (Fig. 2 and 3) must be supplied. This transformation is not trivial, and developers from the PIM layer should not come into contact with its full complexity. In the next section, we propose to split the PSM layer into two parts to simplify the transition from PIM to PSM and enable clearer views on the separate concerns of the PSM layer.

9.4.1 PSM Layer: Grid Profile

The main issue on the PSM layer is the complexity of the UML diagram, if all aspects of the business and Grid logic are modelled in one layer. Fig. 2 and 3 introduced in section 9.3.2 show such a complex view. We propose to divide the PSM layer into two sub-layers, one containing the business view with only a few selected Grid concerns and the other containing the pure Grid view. The business PSM sublayer contains a similar view to the PIM layer but with some Grid specific stereotypes to mark where Grid concerns affect the business logic and Grid wrapper classes which simplify the access to Grid components on the Grid PSM sublayer. The Grid PSM sublayer holds the detailed view of the Grid specific components and is a close representation of what will be implemented in the code layer. This refined MDA architecture is shown in Fig. 6.

9 Model Driven Development of Service-Oriented Grid Applications

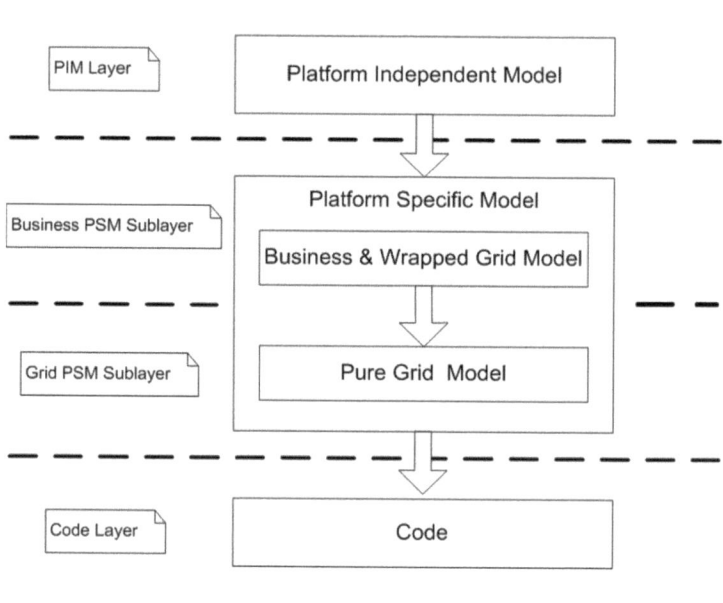

Fig. 6: Refined MDA model

To facilitate the separation of the PSM layer, we introduce an UML Grid Profile to model the Grid concerns in the business layer. Fig. 7 shows the Grid Profile metamodel. The profile consists of three stereotypes: GridClass, GridMethod and GridResourceAttribute. GridClass can be used to mark a class, GridMethod an operation and GridResourceAttribute an attribute. This is similar to the process of marking PIM classes to prepare them for transformation to a PSM suggested by the OMG [12]. But unlike the OMG approach, the developer does not place the marks in an invisible PIM add-on layer, but into the upper PSM layer. We believe this leads to greater clarity, because no invisible components influence the transformations, and the PIM layer stays truly system independent.

9.4 An MDA Approach to Service-Oriented Grid Computing

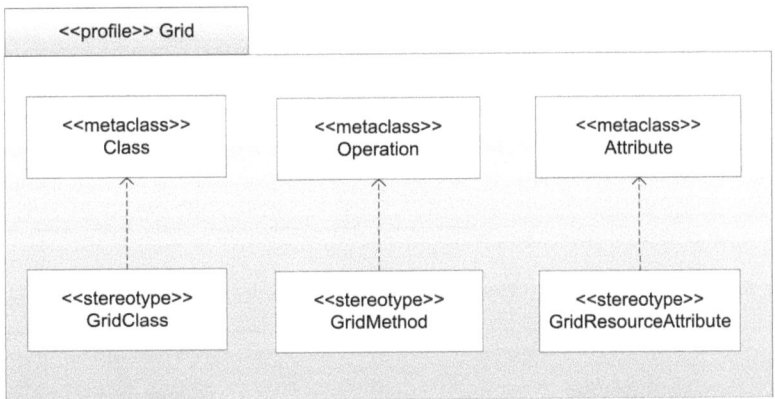

Fig. 7: Grid profile

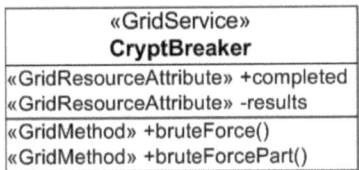

Fig. 8: Annotated service

Fig. 8 shows the business view of the service from Fig. 2 which has been marked with stereotypes from the Grid profile. We also generate wrapper classes which make interacting with the Grid specific classes easier, since only the bare minimum of Grid concerns are exposed there. In this case, we generate a simplified class for the CryptClient which wraps the CryptPortType and CryptFactoryAddressingLocator to access the remote Grid service and thus shields the business logic developer from Grid specific concerns.

All the developers need to do to move from the PIM to both the upper and lower PSMs is to mark which classes, methods and attributes should be exposed via the Grid and specify the target namespace. The transformation tool will then automatically generate all required Grid classes and supporting configuration files.

9 Model Driven Development of Service-Oriented Grid Applications

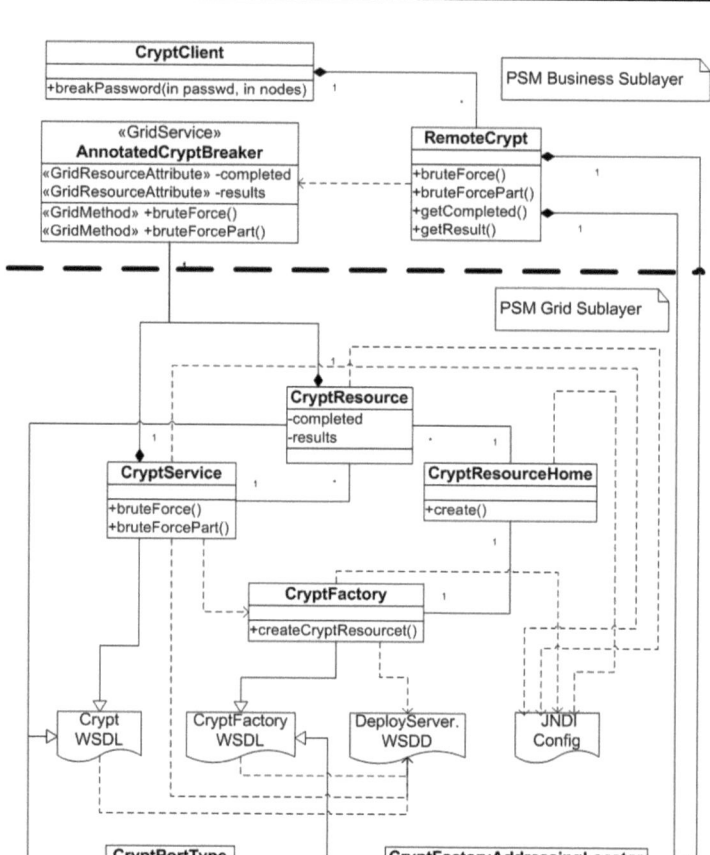

Fig. 9: Grid service UML

Fig. 9 shows both sublayers of the PSM layer separated by a dotted line. The business concerns lie almost entirely in the upper layer and the Grid concerns entirely in the lower layer. The remaining overlap should not be removed from the models, since otherwise it would no longer be possible to understand the connection between the Grid middleware and the business logic.

The automatic generation of the Grid PSM sublayer represents a great benefit for service-oriented Grid computing, since up to now the entire PSM layer had to be created by hand, involving both business logic and Grid developers. Now only the business specific concerns need to be dealt with. The automatic transformation allows non-Grid domain experts to create Grid services simply by marking their PIM UML model with Grid specific stereotypes. This also frees the Grid domain experts, since they do not need to be involved in every step of the business logic development. Any advances in Grid design made by the domain experts can be integrated into the transformers and thus will automatically be integrated into all relevant areas of the business logic. Through the introduction of two sublayers within the PSM layer, we ensure a separation of concerns for business and Grid designers on the PSM layer.

9.4.2 Code Layer: Java Annotations

In the previous section, the PSM layer was divided into a business and a Grid part to facilitate the separation of development concerns. As shown above, the standard Grid service code is quite tangled, forcing developers in the code layer to deal with business and Grid concerns at once. The transformation from PIM to the first and second PSM sublayers already simplifies the code development process somewhat, because the Grid configuration files and support classes can now be transformed into code quite simply. Three tricky cases remain: the service class itself, the service resource class (see Fig. 4) and the client (see Fig. 5). The latter is dealt with by using the generated RemoteCrypt class which wraps the Grid specific CryptPortType and CryptFactoryAddressingLocator. The simplified client is shown in Fig. 10. All the client needs to do is instantiate the RemoteCrypt class with the IP and port number of the node where the service is located. The RemoteCrypt class has the same interface as the CryptBreaker class from the PIM layer and thus allows a Grid free view of the client, freeing the business logic developer from dealing with Grid specific code.

```
String nodeIP = "http://xxx.xxx.xxx.xxx:8080";
RemoteCrypt remoteCrypt = new RemoteCrypt(nodeIP);
remoteCrypt.bruteForce("somePassword");
```

Fig. 10: Simplified client code to call a service

```
@GridClass public class AnnotatedCryptBreaker
{
  @GridResourceAttribute int completed=0;
  @GridResourceAttribute Results results=new Results();
  @GridMethod public String bruteForce(String crypt)
  {
    //for all possible passwords
    //check if password matches
    //developer supplied pure Grid free Java code
  }
  @GridMethod public String bruteForce(String crypt,
    int[] lower, int[] upper)
  {
    //for all possible passwords within upper and lower
    //check if password matches
    //developer supplied pure Grid free Java code
  }
  private boolean matches(String encryptedPassword,
    String enteredPassword)
  {
    //developer supplied pure Grid free Java code
  }
}
```

Fig. 11: Annotated service code

The tangled code problem of the service class and its resource class is solved by using Java annotations. The AnnotatedCryptBreaker class from the upper PSM sublayer can be transformed into a pure Java class by exchanging the Grid Profile stereotypes for Java annotations. Fig. 11 shows this class. The class has been annotated with @GridClass, the methods which need to be exposed to the Grid are annotated with @GridMethod and the class members which comprise the service resource are annotated with @GridResourceAttribute. The actual functionality of the class (i.e. breaking the password) is either created using traditional UML and MDA techniques (not covered here), or by filling in the functionality into the generated classes by hand, as it was done in this example. The functionality of the class consists of pure, Grid code free Java. The class can be compiled and used outside of a Grid environment without any modifications. By exchanging

9.4 An MDA Approach to Service-Oriented Grid Computing

the annotations, the same business logic can be tied into a different platform (e.g. pure web services). To tie the class into the Grid service, custom GridAnnotationProcessors were implemented and can be used to generate Grid service code which redirects calls to the Grid service into the annotated class. The same works for calls to the service resource. By redirecting all calls applicable to the @GridMethods and @GridResourceAttributes from the Grid service and its resource, we free the Grid developers from integrating business logic into their code and we free the business logic developer from dealing with Grid specific code in their business logic.

The only interaction the Grid developer needs to cope with is the redirection of the business method calls. Since this is generated by the AnnotationProcessor, it should not pose any significant burden on the Grid developer. Fig. 12 shows such a redirected call. The method call getAnnotatedService() returns the instance of the annotated class containing the business logic and is the only cross-domain concern the Grid developer is required to deal with. The BruteForceResponse is wrapped and converted back to the pure Java view by the RemoteCrypt wrapper. The same approach is taken for the service resource. The slightly higher overhead of the additional indirection is negligible in comparison to the benefits of cleaner code and fewer cross-domain concerns. To see the code layer benefits, compare Fig. 11 with 4 and Fig. 10 with 5. The generation process itself takes roughly one second on a Pentium M 1.7 GHz notebook computer running MS Windows XP.

```
public BruteForceResponse bruteForce(BruteForce
   complexType) throws RemoteException
{
   AnnotatedCryptBreaker cryptBreaker = getResource()
     .getAnnotatedService();
   BruteForceResponse response = new BruteForceResponse();
   response.setBruteForceReturn(cryptBreaker
     .bruteForce(complexType.getCrypt()));
   return response;
}
```

Fig. 12: Service method redirection

9.5 Conclusions

In this chapter, it was illustrated how a MDA approach can be utilized to deal with the complexity of developing service-oriented Grid applications. A UML Grid Profile was introduced and subsequently a simple Unix password breaking example application was designed on the PIM layer and transformed step by step down to the code layer. To facilitate the MDA transformations introduced, we subdivided the PSM layer into a business logic sublayer and a Grid sublayer. The benefit of this is that the transformations needed to traverse the models from the PIM layer down to the code layer become simpler, requiring less human interaction and at the same time business logic and Grid concerns are separated, allowing each domain expert to work in his or her field. The separation of concerns introduced on the PSM layer is mirrored on the code layer by the use of Java annotations, allowing the same business code to run in different domains simply by exchanging the PSM marks and Java Annotations.

There are several areas for future research. For example, currently only the transformations from PIM to PSM to code are implemented, future work will consist of creating full MDA round-trip support for service-oriented Grid computing. Furthermore, the concept of Grid notifications will be integrated into the Grid Profile and the Java annotation processors to offer more complete support for Grid application development. Finally, the investigation on how more comprehensive real-life Grid applications could benefit from the approach proposed in this chapter is an interesting area of future work.

9.6 Acknowledgements

This work is financially supported by Siemens AG, Corporate Technology, München, by an IBM Faculty Award (Eclipse Innovation Grant) and by the Deutsche Forschungsgemeinschaft (SFB/FK 615, Teilprojekt MT).

9.7 References

[1] S. Andreozzi, P. Ciancarini, D. Montesi, and R. Moretti. Towards a Metamodeling Based Method for Representing and Selecting Grid Services. In *Lecture Notes in Computer Science, Volume 3270*, pp. 78 – 93, 2004.

[2] A. Brown. An Introduction to Model Driven Architecture Part I: MDA and Today's Systems, 2004. IBM The Rational Edge. http://www-106.ibm.com/developerworks/rational/library/3100.html

[3] D. Byrne, A. Hume, and M. Jackson. Grid Services and Microsoft .NET. In *Proc. of UK e-Science All Hands Meeting*, pp. 129–136, 2003.

[4] J. Bézivin, S. Hammoudi, D. Lopes, and F. Jouault. Applying MDA Approach for Web Service Platform. In *Proc. of the 8th IEEE International Enterprise Distributed Object Computing Conference*, pp. 1–13, IEEE Press, 2004.

[5] E. Christensen, F. Curbera, G. Meredith, and S. Weerawarana. Web Services Description Language (WSDL) 1.1, 2001. http://www.w3.org/TR/wsdl.

[6] D. Flater. Impact of Model-Driven Standards. In *Proceedings of the 9th IEEE Conference on System Sciences*, pp. 285–295, IEEE Press, 2002.

[7] F. Flore. MDA: The Proof is in Automating Transformations Between Models. In *OptimalJ White Paper*, pp. 1–4, 2003.

[8] I. Foster, D. Berry, A. Djaoui, A. Grimshaw, B. Horn, H. Kishimoto, F. Maciel, A. Savvy, F. Siebenlist, R. Subramaniam, J. Treadwell, and J. von Reich. The Open Grid Services Architecture, Version 1.0, Whitepaper GGF, pp. 1-19, 2004.

[9] I. Foster, C. Kesselman, J. Nick, and S. Tuecke. The Physiology of the Grid: An Open Grid Services Architecture for Distributed Systems Integration. In *Open Grid Service Infrastructure WG, Global Grid Forum*, pp. 1-31, 2002.

[10] P. Fraternali and P. Paolini. Model-Driven Development of Web Applications: The Autoweb System. In *ACM Transactions on Information Systems, Vol. 28*, pp. 323–382, 2000.

[11] A. Gokhale and B. Natarajan. Composing and Deploying Grid Middleware Web Services Using Model Driven Architecture. In *Lecture Notes in Computer Science, Volume 2519*, pp. 633 – 649, Springer-Verlag, 2002.

[12] J. Mukerji and J. Miller. Overview and Guide to OMG's Architecture, Whitepaper, IBM The Rational Edge, pp. 1-62, 2001.

[13] R. Mulye. Modeling Web Services Using UML/MDA, pp. 1-3, 2005. http://lsdis.cs.uga.edu/ranjit/ academic/essay.pdf.

[14] OASIS. Web Services Resource Framework, 2004. www.oasisopen.org/commitees/tc_home.php?wg_abbrev=wsrf.

[15] O. Patrascoiu. Mapping EDOC to Web Services using YATL. In *Proceedings of the 8th IEEE International Enterprise Distributed Object Computing Conference*, pages 1–12, IEEE Press, 2004.

[16] The Globus Project homepage. Globus Toolkit 4.0 2004. www.globus.org/toolkit

[17] R. Grønmo, D. Skogan, I. Solheim and J. Oldevik, Model-Driven Web Services Development. In *Proc. of the 2004 IEEE Int. Conference on e-Technology, e-Commerce and e-Service (EEE'04)*, pp. 633 – 649, IEEE Press, 2004.

[18] R. Radhakrishnan and M. Wookey. Model Driven Architecture Enabling Service Oriented Architectures. Whitepaper Sun Microsystems, pp. 1 – 13, 2004.

[19] D. Skogan, R. Gronmo, and I. Solheim. Web Service Composition in UML. In *Proceedings of the 8th IEEE International Enterprise Distributed Object Computing Conference*, pp. 1–11, IEEE Press, 2004.

[20] M. Smith, T. Friese, and B. Freisleben. Towards a Service-Oriented Ad Hoc Grid. In *Proceeedings of the 3rd International Symposium on Parallel and Distributed Computing, Cork, Ireland*, pp. 201–209, IEEE Press, 2004.

[21] The World Wide Web Consortium. Simple Object Access Protocol (SOAP), 2003. http://www.w3.org/TR/soap/.

[22] D. Thomas. The Impedance Imperative: Tuples + Objects + Infosets = Too Much Stuff! In *Journal of Object Technology*, pp. 1–6, 2003.

Index

A

Abrechnung • 22, 28, 34
Abrechnungsmodell • 13, 14
Accounting • 22, 32, 77, 152, 186
ad hoc Grid • 144, 146
allocation • 64, 75, 106, 150, 151
Allocation • 62
allocative efficiency • 76
allocator • 67
Allokation • 19, 39, 41, 58
Allokationseffizienz • 38, 54
Allokationskosten • 42
Allokationsrate • 38, 52, 54
Anbahnungsphase • 36
Anwendung • 137
Anwendungsschicht • 5
Anwendungsschichtnetz • 52
Application Service Providing, 2. Generation (ASP-2) • 28
Application-Grid • 22
auction mechanism • 71
Authentisierung • 22, 27
autonomic computing • 186
Autorisierung • 22, 27

B

Bag of Tasks (BoT) • 151
Basisdienste • 27
bid • 66
Billing • 32, 152, 186
billing system • 150
black box • 67, 133, 134, 135
Broadcast • 154, 156
Broker • 42
Broker-Komponente • 19

bundle exchanges • 75

C

C • 122, 123, 185
C++ • 122, 185
Choreographie • 23
Clash-Analyse • 3
Cluster • 8, 73, 85, 106, 107, 114, 121
Cluster Management • 8
Cluster-Infrastruktur • 13
Cluster-System • 7, 13, 137
Cluster-Technologie • 9
Codine • 2
Collaborative Engineering • 6
collaborative projects • 107, 124
commodity markets • 73
computational resources • 64, 75, 145
computational tractability • 69
compute clusters • 107, 109
Compute-Grid • 21
Computer Aided Engineering (CAE) • 4, 7
confidentiality • 20, 86, 92
Content Distribution Netzwerk (CDN) • 40, 50
Cooperative Bartering and Sharing • 73
Corba • 193, 194

D

data management • 86, 94
Data-Federation • 21
Data-Grid • 21
Datenschutzprobleme • 35
Datensicherheit • 141
Designparameter • 137
dezentral • 9, 22, 45, 137
Dienstauswahl • 37, 41, 58

211

Dienstauswahlmethode • 58
Dienstauswahlverfahren • 39, 48
Dienstfindung • 41, 54
Dienstsuche • 36
Dienstzugriffsmodell • 34
Double-sided mechanism • 70
Dutch Combinatorial Auctions • 74

E

eBusiness • 17
eBusiness Umgebung • 24
eCommerce • 18
eEngineering • 18
electronic markets • 68
Electronic Markets • 83
encryption • 162
enhanced Business • 18
eScience • 18, 30
Evaluation • 100, 194
Externalisierung • 36

F

fault management • 106
Fehlallokation • 44
Financial Services-Sektor (FSS) • 7
firewalls • 146
FORTRAN • 185
Funktionssicherheit • 27

G

Generalized Vickrey Auction • 72, 74
Gesch ftslogik • 8
Gesch ftsmodell • 17, 24, 34
Gesch ftsprozesse • 5, 17, 27
Gie prozess • 135, 137
Gie simulation • 133
Global Grid Forum (GGF) • 24
Globalisierung • 29
GRACE • 47

granularity • 114, 124
Grid market • 64, 75
Grid Middleware • 11, 123, 145, 150, 192, 204
Grid service • 24, 123, 124, 144, 194, 205
Grid Service archive • 163, 177
Grid Technologie • 1, 17, 65, 134, 178
Grid-Architektur • 7
Grid-basierter Kooperationswerkzeuge • 138
Grid-Cluster • 7, 9
Grid-Infrastrukturen • 11
Grid-Nutzer • 23
Grid-Provider • 22
Grid-Verbund • 137

H

heterogeneity • 86, 144
hot deployment • 144, 163, 175

I

idle resources • 66
incentive compatibility • 68, 69, 76
Independent Software Vendor (ISV) • 6
Informationsdienste • 31, 32
Insourcing • 35
Instant Messaging • 40
Integration • 12, 24, 29, 87, 92, 99, 151, 191
Integrationsaufwand • 10
interaktiv • 138
interconnected workstations • 106
intermediaries • 67
Internet • 18, 40, 46, 107, 143, 163
internet applications • 85
internet infrastructure • 146, 153
Internet Service-Provider • 31
internet standards • 146
internet technologies • 85
ITC-Dienstleister • 31
ITC-Leistungen • 18, 19
ITC-Services • 17, 20, 29
IT-Dienste • 34

IT-Dienstleistungen • 12, 33, 34, 58
Iterative Combinatorial Auctions • 74
IT-Infrastrukturen • 12, 134

J

J2EE • 193
Java • 122, 150, 160, 194, 205
Java Provider • 166

K

Kalibrierung • 137, 138
katallaktisch • 47, 58
Katallaxie • 47
kleine und mittelständische Unternehmen (KMU) • 3, 35, 133
Kommunikationsinfrastrukturen • 40
Konvergenz • 21
Konzept • 15, 23, 33
Koordinator • 42, 44, 54
koordinatorfrei • 33
Kosten • 6, 34, 35, 47
Kostenersparnis • 2, 33

L

Lastverteilung • 2, 11
legacy code • 147
Life Science Grid • 107
Life Sciences • 7, 99, 107
Lizenz • 28, 31
Lizenzkosten • 12
load balancing • 106

M

Market Engineering • 68
market mechanism • 64, 73, 76
market participant • 66, 67
Markov Chain • 113, 125

Maßberschneidungsanalyse • 3
Messaging-Dienste • 31
metacomputing • 85
Metrik • 28, 33, 38, 48
Middleware • 2, 106, 112, 124, 147, 151, 192
middleware developers • 192
Middleware-Anbieter • 7
mittelständische Unternehmen • 35, 133
model driven architecture (MDA) • 192
model sweep • 121
model-guided discovery • 105, 108
Monte Carlo method • 97, 105, 112
Monte Carlo-Simulation • 7, 9, 109
multi-application • 152
Multicast • 40, 156
multi-disciplinary • 152
multi-organizational • 152

N

neighborhood sweep • 121
Network Operation System • 11
Netzbetreiber • 24
Netzwerkszenarien • 33
Netzwerktopologie • 45, 51
non-intrusive • 145, 148, 165

O

Object Request Broker (ORB) • 36
Offshoring • 29
On-demand Computing • 33, 58, 79
Open Agoric Systems • 47
Open Grid Service Architecture (OGSA) • 24
Operating System • 123, 147, 150, 157, 192
Optimierung • 47, 136, 137
organizational boundaries • 65, 79, 150
Outsourcing • 20, 30
Outsourcing-Entscheidung • 36
overlay networks • 153, 162
Overlaynetz • 58

P

parallel computing • 85
Parallelrechner • 135, 140
parameter fitting • 97, 103, 105, 109, 112, 114, 124, 125
Patch-Management • 8
Peer-to-Peer (P2P) • 40, 45, 50, 65, 150, 153, 159
Platform Independent Model (PIM) • 192
Platform Specific Model (PSM) • 193
Preismodell • 29
Problemdefinition • 138
Probleml sungsumgebung • 24
process chain • 86
process monitoring • 107
Produktionsprozess • 134, 136
Proxies • 28
Prozesssimulation • 138
Public Key Infrastructure (PKI) • 13

Q

Quality of Service (QoS) • 179

R

Rechenkapazit t • 21
Rechenleistung • 2, 21, 33, 40, 59, 137, 139
Rechnerverbund • 137
Remote Procedure Call (RPC) • 26
Repeated Vickrey Auction • 72
replication • 73, 155
Replikation • 6
Resource Broker • 41, 52, 67
resource consumer • 67, 70
resource information manager • 67
Resource Management Framework (RMF) • 154
resource sharing • 64, 65, 85, 150
Resource-Provider • 22, 73, 78
Ressource-Grid • 22
Ressourcenallokation • 58

Rubens Cluster • 124, 141

S

safety • 27
sandbox • 145, 169, 173
scalability • 150, 153
scalable • 105, 112
scheduling • 65, 72, 150, 181
Schichtenmodell • 24
Schnittstellen • 4, 22, 27, 28
Security • 27, 73, 86, 144, 148, 168, 178
separation of concerns • 193, 195, 196, 205, 208
service deployment • 156, 167, 185
service discovery • 144, 147, 153, 158, 159, 183, 185
Service Grid-Umgebung • 27
Service Level • 9, 15, 19
Service Level Agreement (SLA) • 5, 73, 185
Service-Broker • 25
Service-Grid • 17, 23, 32
Servicekomponenten • 29
service-oriented architecture (SOA) • 191
Serviceorientierte Architektur (SOA) • 31, 36, 49, 191, 195
Service-Provider • 29, 31, 159, 178, 180
Shared Services • 19, 27
Sicherheit • 13, 27, 32, 35, 134, 139
Sicherheitsarchitektur • 26
Sicherheitsdom ne • 22
Sicherheitsstandard • 20
Sicherheitstechnik • 28
Simple Object Access Protocol (SOAP) • 26, 36, 143, 152, 159, 191
Simulation • 18, 31, 47, 58, 73, 78, 87, 105, 133, 138
simulation algorithm • 112
simulation code • 112, 118
Simulation Data and Process Management (SDM) • 6
simulation problems • 85
simulation process • 87
simulation specialist • 108

simulation step • 105, 111, 121
simulation tools • 93, 107
simulation workbench • 91
simulation workflow • 89
simulation-based optimization • 109
Simulationsdaten-Management System • 6
Simulationslauf • 49
Simulationsmodell • 138
Simulationsmodellierung • 52
Simulationsrechnung • 2
Simulationsspezialist • 137
Simulationsumgebung • 48, 51, 136
Single Sign-On • 22
Skalierbarkeit • 47, 58
SLA-Verletzungen • 19
SOA-Konzept • 31
Softwareagenten • 47, 52
Spezifit t • 36
Spitzenbelastung • 9
Standard Vickrey Auction • 72
stochastic simulation • 76
Supercomputer • 124
Supercomputerzentren • 2
Supercomputing • 90, 93
Supercomputing Center • 1
supercomputing centre • 85, 91
System Provisioning • 11
System-Integratoren • 12
Systemmanagement • 8, 11
Systems Biology • 97
Systems Biology Markup Language (SBML) • 127
Szenarien • 3, 134, 139
Szenario • 2, 50, 135

T

Technologie • 7, 59, 85, 101, 156, 193
Transaktion • 42, 45, 54
Transaktionskosten • 36, 37
Transaktionspartner • 41
Transaktionsphasen • 35, 36, 37
trust • 145, 178, 185
tunnelling • 159

U

Ubiquitous Computing • 50
Uniform Access to Computing Resources (UNICORE) • 22, 31, 152
Universal Description, Discovery and Integration (UDDI) • 26, 36, 158

V

Value at Risk • 7
verteilte Anwendung • 3
Vertrauen • 28
Vertraulichkeit • 20, 28
virtual prototyping • 93
Virtualisierung • 11, 15, 22, 25, 34
virtualization • 147, 191
Virtualization • 15, 78
Virtualization Engine • 15
virtuelle Fabrik • 134, 136
virtuelle Organisation • 6, 20, 22
virtueller Computer • 34
visualization • 86, 90, 93, 108, 118, 122

W

web portal • 123
Web Service Notification • 37
Web Service-Resource Framework (WSRF) • 10, 37
Web Services Definition Language (WSDL) • 25, 36, 143, 152, 158, 162, 195, 199
Web Technologie • 21
web-portal • 124
Workflow • 26, 29, 86, 87, 92, 191
workflow engine • 151, 183
workflow engines • 186
Workstation-Grids • 7
Workstations • 2, 7, 72, 85, 91, 94, 106, 144, 145, 150
World Wide Web • 20, 46, 159

Z

zentral • 4, 42, 43, 141

Zentralit t • 41
Zulieferer • 3, 133
Zwischenschicht • 18

Mit Bestsellern aus dem Bereich IT lernen

Dietrich May
Grundkurs Software-Entwicklung mit C++
Praxisorientierte Einführung mit Beispielen und Aufgaben - Exzellente Didaktik und Übersicht
2., überarb. u. erw. Aufl. 2006. XVI, 540 S. Br. € 29,90   ISBN 3-8348-0125-9
Programmentwicklung von der Idee bis zur sicheren Lösungsstrategie - Darstellung von Zahlen und Zeichen, Logik, Dateiaufbau - Grundlagen der Programmier-Elemente: Schleifen, Wiederholungen, Funktionen, Klassen, Objekte, Vererbung - Methoden zur systematischen Software-Entwicklung

Dietmar Abts
Grundkurs JAVA
Von den Grundlagen bis zu Datenbank- und Netzanwendungen
4., verb. u. erw. Aufl. 2004. X, 408 S. mit Online-Service. Br. € 19,90
ISBN 3-528-35711-8
Klassen, Objekte, Interfaces und Pakete - Ein- und Ausgabe - Multithreading - Grafische Oberflächen mit Swing, Applets - Datenbankzugriffe mit JDBC - Netzanwendungen - Spracherweiterungen der Version J2SE 5.0

André Maassen/Markus Schoenen/Ina Werr
Grundkurs SAP R/3®
Lern- und Arbeitsbuch mit durchgehendem Fallbeispiel - Konzepte, Vorgehensweisen und Zusammenhänge mit Geschäftsprozessen
3., durchges. u. verb. Aufl. 2005. XXIV, 608 S. mit 256 Abb. u. 25 Tab. Br. € 39,90   ISBN 3-528-25790-3
Technische Aspekte - Benutzerkonzept und Handhabung - Unternehmensstrukturen in Personalwirtschaft, Materialwirtschaft, Vertrieb und Finanzwesen - Fallstudiengestützte Einführung in die Arbeit mit Stammdaten und Bewegungsdaten - Screenshot-basierte Arbeitsanweisungen - Erste Schritte mit Report Painter und Report Writer

vieweg

Abraham-Lincoln-Straße 46
65189 Wiesbaden
Fax 0611.7878-400
www.vieweg.de

Stand 1.1.2006. Änderungen vorbehalten.
Erhältlich im Buchhandel oder im Verlag.

Grundlagen verstehen und umsetzen

Andreas Gadatsch
Grundkurs Geschäftsprozess-Management
Methoden und Werkzeuge für die IT-Praxis:
Eine Einführung für Studenten und Praktiker
4., verb. u. erw. Aufl. 2005. XXIV, 460 S. mit 335 Abb. Br. € 34,90
ISBN 3-8348-0039-2

Gunther Friedl/Christian Hilz/Burkhard Pedell
Controlling mit SAP®
Eine praxisorientierte Einführung - Umfassende Fallstudie -
Beispielhafte Anwendungen
4., verb. u. erw. Aufl. 2005. XXII, 275 S. Br. € 39,90   ISBN 3-8348-0101-1
Überblick über Controlling mit SAP - Durchgängige Fallstudie - Kostenstellenrechnung - Produktkalkulation und Kostenträgerrechnung - Ergebnis- und Marktsegmentrechnung - Konzeptionelle Entwicklungen des Controlling und ihre Abdeckung durch SAP (SEM, BW) - Vorbereitende Tätigkeiten im Customizing - Nutzung von Vorlagemandanten

Paul Alpar/Heinz Lothar Grob/Peter Weimann/Robert Winter
Anwendungsorientierte Wirtschaftsinformatik
Strategische Planung, Entwicklung und Nutzung von Informations- und Kommunikationssystemen
4., verb. u. erw. Aufl. 2005. XVI, 495 S. mit 199 Abb. u. Online Service.
Br. € 29,90   ISBN 3-528-35656-1
Informations- und Kommunikationssysteme in Unternehmen - Informations- und Wissensmanagement - Controlling der Informationsverarbeitung - Ganzheitliche Gestaltung von Informations- und Kommunikationssystemen - Architektur betrieblicher Anwendungssysteme - Methoden und Werkzeuge zur Entwicklung und Einführung von Software - Informations- und Kommunikationstechnologie

Abraham-Lincoln-Straße 46
65189 Wiesbaden
Fax 0611.7878-400   Stand 1.1.2006. Änderungen vorbehalten.
www.vieweg.de   Erhältlich im Buchhandel oder im Verlag.

Masterkurse

Alfred Nischwitz, Peter Haberäcker
Masterkurs Computergrafik und Bildverarbeitung
Alles für Studium und Praxis - Bildverarbeitungswerkzeuge,
Beispiel-Software und interaktive Vorlesungen online verfügbar
2004. XXXII, 860 S. mit Farbteil und Online-Service. Br. € 49,90
ISBN 3-528-05874-9

Andreas Gadatsch, Elmar Mayer
Masterkurs IT-Controlling
Grundlagen und Strategischer Stellenwert -
IT-Kosten- und Leistungsrechnung in der Praxis -
Mit Deckungsbeitrags- und Prozesskostenrechnung
2., verb. u. erw. Aufl. 2005. XXII, 502 S. mit Online-Service. Br. € 49,90
ISBN 3-528-15849-2

Das Leitbildcontrolling-Konzept für die IT - IT-Controlling: Vom Konzept zur
Umsetzung (Zielformulierung, Zielsteuerung, Zielerfüllung) - Einsatz strategischer IT-Controlling-Werkzeuge - Operative Werkzeuge - IT-Kostenrechnung

Franz Klenger/Ellen Falk-Kalms
Masterkurs Kostenstellenrechnung mit SAP®
R/3® Enterprise - Mit Testbeispiel und Customizing - Für Studenten und
Praktiker
4., verb. u. erw. Aufl. 2005. XVI, 441 S. mit 141 Abb. u. Online Service.
Br. € 29,90   ISBN 3-3-8348-0026-0

Organisationsstruktur und Stammdaten in Finanzbuchhaltung und Kostenrechnung - Planung von Primärkosten, innerbetrieblicher Leistungsverrechnung und Kalkulationsparametern - Istdatenerfassung - Soll-/Ist-Vergleich - Interaktives Berichtswesen - Schnittstelle zur ABAP-Programmierung - Tipps zum Einsatz von Standardsoftware in der Lehre

Abraham-Lincoln-Straße 46
65189 Wiesbaden
Fax 0611.7878-400   Stand 1.1.2006. Änderungen vorbehalten.
www.vieweg.de   Erhältlich im Buchhandel oder im Verlag.

If you have any concerns about our products,
you can contact us on
ProductSafety@springernature.com

In case Publisher is established outside the EU,
the EU authorized representative is:
**Springer Nature Customer Service Center GmbH
Europaplatz 3, 69115 Heidelberg, Germany**

Printed by Libri Plureos GmbH
in Hamburg, Germany